The John Dewey Essays in Philosophy

Edited by the Department of Philosophy, Columbia University, and including the John Dewey Lectures

Number Three

Teleology Revisited and Other Essays in the Philosophy and History of Science

THE JOHN DEWEY ESSAYS IN PHILOSOPHY

W. V. Quine, *Ontological Relativity and Other Essays*, 1969

A. J. Ayer, *Probability and Evidence*, 1972

Ernest Nagel, *Teleology Revisited and Other Essays in the Philosophy and History of Science*, 1979

Teleology Revisited and Other Essays in the Philosophy and History of Science

Ernest Nagel

Columbia University Press New York 1979

This volume consists of the third of the John Dewey Lectures delivered at Columbia University under the auspices of the Department of Philosophy and eleven additional essays. Ernest Nagel was John Dewey Professor of Philosophy at Columbia from 1955 to 1966. He is now University Professor Emeritus.

Library of Congress Cataloging in Publication Data

Nagel, Ernest, 1901–
Teleology revisited and other essays in the philosophy and history of science.

(The John Dewey essays in philosophy; no. 3)
Includes bibliographical references and index.
1. Science—Philosophy—Addresses, essays, lectures. 2. Logic—Addresses, essays, lectures. 3. Teleology—Addresses, essays, lectures. I. Title. II. Series.
B67.N33 501 78-14037
ISBN 0-231-04504-2

Columbia University Press
New York—Guildford, Surrey

Printed in the United States of America

To the memory of

IRENE and WILLIAM FINKEL

Acknowledgments

All the essays in this volume have been previously published substantially in their present form, and I wish to thank their publishers and the holders of copyright to them for permission to reprint.

"Modern Science in Philosophical Perspective" was presented at the New York Academy of Medicine in the series *Lectures to the Laity, Nos.* XIX/XX, under the title "The Philosopher Looks at Science," and was published in a slightly shorter form in the volume *Medicine and the Other Disciplines,* ed. Iago Galdston, by International Universities Press, New York, copyright © 1959.

"Theory and Observation" was presented at The Johns Hopkins University in 1969 in the series The Alvin and Fanny Blaustein Thalheimer Lectures, and was published in the volume Ernest Nagel, Sylvain Bromberger, Adolf Grünbaum, *Observation and Theory in Science,* by The Johns Hopkins Press, Baltimore, copyright © 1971.

"Some Reflections on the Use of Language in the Natural Sciences" was read at a Symposium on "The Theory of Language" at the Conference on Methods in Philosophy and the Sciences, New School for Social Research, New York City, May 6, 1945, and published in the *Journal of Philosophy,* Vol. XLII, 1945.

"The Quest for Certainty" was published in *Sidney Hook and the Contemporary World,* ed. Paul Kurtz, by The John Day Company, copyright © 1968 (now owned by Thomas Y. Crowell Company, New York).

"Philosophical Depreciations of Scientific Method" first appeared in *The Humanist,* July/August 1976 and is reprinted by permission.

"Issues in the Logic of Reductive Explanations" was read at *The International Philosophy Year Conferences at Brockport* in 1968, and was published in the volume *Mind, Science, and History,* ed. H. E. Kiefer and M. K. Munits, by the State University of New York Press, Albany, copyright © 1970.

"Carnap's Theory of Induction" was published in Paul A. Schilpp, ed., *The Philosophy of Rudolf Carnap,* THE LIBRARY OF LIVING PHILOSOPHERS, by the Open Court Publishing Company, La Salle, Illinois, copyright © 1963.

" 'Impossible Numbers': A Chapter in the History of Modern Logic" first appeared in *Studies in the History of Ideas,* edited by the Department of Philosophy of Columbia University, Vol. III, published by Columbia University Press, New York, 1935.

"The Formation of Modern Conceptions of Formal Logic in the Development of Geometry" was published in *Osiris,* Vol. 7, 1939.

"Determinism and Development" was read at a Conference on Child Growth and Development held at the University of Minnesota, December

8–10, 1955, and published by the University of Minnesota Press, Minneapolis, copyright © 1957.

"Some Notes on Determinism" is taken from the proceedings of the third annual New York University Institute of Philosophy held at Washington Square, New York, May 15–16, 1959, and was published in *Determinism and Freedom*, ed. Sidney Hook, New York University Press, copyright © 1960.

The two-part final essay from which the book takes its title consists of the third series of John Dewey Lectures delivered at Columbia University on March 28 and 29, 1977, and was published in the *Journal of Philosophy*, Vol. LXXIV, 1977. I want to express my deep appreciation of the honor the Columbia Department of Philosophy bestowed upon me in inviting me to be the third John Dewey Lecturer, and also to thank the Department for its singular patience in permitting me to delay the delivery of the Lectures for two years.

Contents

Introduction

The essays in this collection were written at different times and for different occasions, and they do not have the close-knit unity of a well-constructed argument for a major conclusion. Nevertheless, in one way or another most of them deal with a common theme and express a common outlook. The theme is the logic of scientific inquiry and some of its alleged limitations. The outlook is that of an unreconstructed empirical rationalist, who continues to believe that the logical methods of the modern natural sciences are the most effective instruments men have yet devised for acquiring reliable knowledge of the world and for distinguishing warranted claims to such knowledge from those that are not.

This belief was once widely held by practicing scientists and commentators on the sciences who in other respects differed in their philosophical commitments. "Given the method of science," Condorcet declared, "one can predict that, other things being equal, knowledge will grow steadily more comprehensive and reliable." And a century later, C. S. Peirce argued in a similar vein for the superiority of scientific method over other ways of settling doubt; and with the former in mind he maintained that "The genius of a man's logical method should be loved and revered as his bride, whom he has chosen from all the world."

However, despite the remarkable experimental and theoretical achievements of the physical and biological sciences during the present century, this belief has been vigorously attacked by, among others, some distinguished scientists. Although Max Born was a major contributor to the development of quantum mechanics, he confessed in the preface to a collection of his semipopular papers[1] that while in 1921 the "scientific method seemed to me superior to other, more subjective ways of forming a picture of the world," thirty years later he "believed in none of these things. . . . I now regard my former belief in the superiority of science over other forms of thought and behavior as a self-deception. . . ." One important reason for Born's change of heart was undoubtedly the fact that the generally accepted interpretation of quantum theory—an interpretation which he had himself proposed—allegedly precludes "an objective knowledge of the world." This is also the reason why other physical scientists (such as the organic chemist James B. Conant, who later became the president of Harvard University) came to deny that "science is an exploration of the universe." More specifically, Conant rejected as false the notion that physicists discover the structural laws of the material world (such as the "truth about the nature of heat, light, and matter"). According to him, scientific theories are not descriptions of the constitution of things, in any way analogous to the way ordinary maps represent "real rivers, mountains, trees, bays" and the like. On the contrary, the significance of scientific theories produced in the past 350 years was said by him to be "the same as the significance of the great periods in history, or the significance of the works of the musical composers."[2]

But the much advertised "paradoxes" that are entailed by the standard interpretation of the mathematical formalism of quantum theory are not the sole reason for current doubts about science as a cognitive enterprise. Nor have physical scientists been alone in raising such doubts. During the past quarter of a century there has been a many-pronged attack on the claim that the logic of scientific inquiry is the paramount instrument for achieving intellectual mastery of nature. Not only natural scientists, but also professional philosophers, historians and sociologists of science, and humanistically oriented social critics have produced a variety of arguments for doubting that "objective" and soundly based knowledge is really achieved in theoretical science; that scientific theories are ever accepted or rejected on the basis of a "rational" evaluation of the available evidence for them; and indeed that there actually is anything like an identifiable "scientific method." Supplying reasons for these sceptical doubts, together with proposing prescriptions for what is rather oddly designated as "the growth of knowledge," is a conspicuous feature of the so-called "new philosophy of science."

The majority of the essays in the present collection are concerned with some of the broad issues, whether explicitly or only indirectly, raised by this "new philosophy of science." These essays seek to show that the latter's scepticism of the efficacy of scientific method for attaining genuine knowledge is not well founded; and it would be pointless to repeat here the arguments they advance for this conclusion. Much of the animus of the "new orientation" in the philosophy of science is directed against the alleged ahistorical character of the "orthodox approach"; against the latter's supposed claim that the observational evidence for a scientific theory can be assessed by using the rules of a formal calculus; and against the view attributed to the "orthodox approach" that the "rationality" of the natural sciences consists in their evaluating evidence in conformity with such rules. These characterizations of the "old philosophy of science" are conceivably true of *some* philosophers in this category (for example, Rudolf Carnap). These characterizations are a caricature of *most* of the older generation of writers on the subject (for example, C. S. Peirce, Josiah Royce, John Dewey, M. R. Cohen, or P. W. Bridgman); and they are not even true of some of Carnap's fellow logical positivists (such as Philipp Frank or Richard von Mises). Unlike Carnap, none of these thinkers subscribed to an ahistorical evaluation of the evidence for a scientific theory; and none of them identified the rationality of science with the use of exclusively formal canons for assessing claims to knowledge. It is misleading to ascribe to all representatives of the "orthodox approach" to the philosophy of science the beliefs that are idiosyncratic of what at best is a relatively small subset of that group of thinkers. In any case, one essay in this collection calls attention to what I still believe, despite Carnap's rejoinder to it, are serious difficulties in his formal logic of inductive probability.

This is not the place to try to do justice to the complex questions that require attention in an adequate analysis of the notion of rationality in science. But it may be helpful for understanding the present discussion to sketch briefly, omitting all complicating qualifications, the view taken of this notion by one tradition associated with the "old philosophy of science." According to this view, it is a fundamental error to define the rationality of a pattern of thought (or more generally of a mode of behavior) in terms of supposedly self-evident *a priori* principles. On the contrary, on this view a mode of behavior is rational if it regularly (or for the most part) achieves the objectives for the attainment of which the behavior is undertaken. In consequence, if the aim of scientific inquiry is to obtain satisfactory answers to the questions (or problems) that initiated the inquiry, the rationality of science consists in its use of procedures (whether symbolical or overtly experimental) which have been shown to be capable of doing just that. There is no antecedent guarantee that a proposed rule for the conduct of inquiry, if adopted, will achieve the aim of the undertaking, however "inherently rational" the rule may seem. Just as the effective rules or modes of procedure for constructing enduring chairs and tables can be ascertained only by reflecting upon the practice of cabinet makers, so the effectiveness of a proposed method for achieving the objectives of inquiry can be established only by reflecting upon the historical fortunes of that method.[3]

In my judgment, and in agreement with this view, a purely formal approach to the task of evaluating claims to knowledge is certainly mistaken. But so is the proposal to exclude *all* formal considerations in the performance of this task. Proponents of the "new philosophy of science" have repeatedly, and I think correctly, claimed that "orthodox" philosophers of science have failed egregiously to construct a calculus, with the help of which it could be shown that the methods of science are effective instruments for arriving at the truth (or even close approximations to the truth). However, it does not follow from this, as some of those proponents have also maintained recently, that such effectiveness of scientific method cannot be shown *at all,* or that in evaluating the evidence for a theory formal relations between evidential statements and theory can be completely ignored.

Two historical papers in this collection are on the face of it unrelated to the questions that are of central concern to the "new philosophy of science." These papers describe certain "internal" developments in algebra and geometry which contributed to the clarification of the nature of mathematics and of its method, as well as to the renascence of formal logic.[4]

These historical studies make no attempt to account for the changes they describe in terms of so-called "external" factors, such as psychological dispositions, economic and political pressures, or social needs. Causal explanations of this sort of episodes in the history of ideas fall, by definition, into the province of the sociology of knowledge (and more specifically into the domain of the sociology of science); but such accounts of scientific change have also been proposed by proponents of the "new philosophy of science."

Sociological explanations of scientific change cannot be dismissed on *a priori* grounds as inherently impossible, and they must be evaluated in the light of the available evidence for them. But neither can it be assumed without further ado, as some sociologists of science have in fact assumed, that there *must* be "external" sociological factors that account for (or "determine"), even if only in part, the content and the acceptance or rejection of scientific theories.[5] There is, to be sure, compelling evidence, much of it gathered by sociologists of knowledge, that there are institutional prerequisites for a flourishing science of the sort currently found in advanced industrial societies, even if those necessary conditions for the existence of modern science have thus far not been exhaustively specified. On the other hand, if *one* aim of the sociology of science is to explain, in terms of "external" factors like those mentioned above, the *genesis and content* of scientific ideas, the *direction* of their development, and their *acceptance or rejection* by the scientific community, then I am confident that in this respect and at present the sociology of science is at best a *program of research,* not a corpus of warranted conclusions.

One small bit of evidence in support of this judgment may perhaps make it seem less dogmatic. In a lengthy article acclaimed by many historians and sociologists of science as an important study,[6] Paul Forman proposed an explanation of why German physicists and mathematicians in the 1920s accepted so readily the new quantum theory with its assumption of a radical indeterminism in nature. According to him, intellectuals in Germany were vehemently antagonistic to the world view, for which they held classical physics responsible, that all processes in the universe are subject "to the rigid, dead hand of the law of causality." By accepting the indeterministic quantum theory, so Forman's explanation runs, German physicists and mathematicians were adapting themselves to a hostile intellectual environment. In other words, he goes on to say, "It was only as and when [the] romantic reaction against exact science had achieved sufficient popularity inside and outside the university to seriously undermine the social standing of the physicists and mathematicians that they were impelled to come to terms with it."[7]

This explanation is perhaps surprising as well as entertaining. But in my opinion the evidence Forman adduces for it is quite insufficient to support his story. I can only sketch my reasons for this judgment. In the first place, it is extremely doubtful on the evidence that the new quantum theory was in fact quickly accepted by German scientists; and if the doubt is justified, as I think it is, Forman's proposed explanation does not explain anything that actually happened. Secondly, there is little solid evidence to show that the "romantic reaction to exact science" was undermining the social standing of those German physicists who continued to subscribe to the deterministic assumptions of classical physics. Nor is there any real evidence to support the claim that in general German scientists accepted the new quantum theory *in order to* accomodate themselves to popular winds of doctrine. And finally, Forman does not take seriously the obvious hypothesis (for which there *is* substantial evidence) that, like their professional colleagues in other lands, German physicists eventually accepted the new quantum theory—even when they disliked its indeterministic assumptions—because the problems and the data *internal* to their discipline gave them no viable alternative. The lack of cogent evidence for Forman's explanation is discouragingly similar to the available evidence for all other sociological explanations of scientific change with which I am familiar. For this reason I do not think that I need apologize for the fact that the two studies in the history of ideas offer no sociological explanations for changes in mathematical doctrine and ignore all "external" causal factors.

The final three essays in this collection deal with several substantive concepts that have important uses in the natural sciences. Two of them are fairly brief. In fact, however, all three of these essays, though they are self-contained, supplement my discussions of those concepts in earlier publications; and I hope that the concluding essay in particular clarifies and improves upon the analyses of the notion of teleology (as it occurs in biology) which I published some time ago. These three essays are not directly related to the issues raised by the "new philosophy of science." But they do illustrate how one major task of the "orthodox" philosophy of science—the task of analyzing substantive scientific notions—may be pursued. This task seems to go out of fashion when the doubtful view is adopted that it is feasible to create useful recipes for the growth of knowledge, and that the central if not exclusive objectives of the philosophy of science is to develop and evaluate such prescriptions.

1/Modern Science in Philosophical Perspective

There is an ancient and still influential tradition according to which philosophy is the supreme architectonic science, competent to prescribe to all inferior disciplines the limits of their possible scope and the ultimate significance of their discoveries. It is not in the spirit of this tradition that the following reflections are offered. Philosophy, as I conceive it, has no special avenue to truth. Whatever knowledge it claims to have of the world must be based, if the claim is warranted, either on the findings of the special sciences or on the same sort of evidence which the special sciences themselves acknowledge. Philosophy is not a rival of science. It neither supplies a more adequate account of things than the sciences provide, nor does it lay bare the alleged "presuppositions" upon which the special sciences supposedly rest, but which they are incompetent to justify with the help of their own methods.

Philosophy is in part a commentary on science. Its task in this regard is two-fold. It aims to analyze the complex logical procedures by which the apparently remote and frequently strange objects of theoretical science are brought into relation with the concrete occurrences of common experience, to specify the functions in inquiry which the abstract notions of scientific thought perform, and to make evident the conditions which must be instituted if warranted claims to knowledge are to be achieved. Secondly, it seeks to make explicit the import, for traditional schemata of morals and for habitual patterns of conduct, of fresh scientific discoveries and of the logical method with which the sciences operate. It is in this way that philosophy becomes a clarified expression of the broad significance of science for human life.

It is my special aim to present, not an exhaustive survey of current philosophies of science, but a brief characterization of what seem to me the dominant features of the intellectual method of science. It is my hope to indicate in this way the humane values implicit in the use and extension of that method, and to place in proper perspective current trends which seek in one way or another to limit the authority and scope of scientific reason.

1

There are at least three distinguishable facets that science presents to a philosophical student. The most obvious one, and the one by far the best advertised, is the practical control over nature which science yields. It would be tedious to rehearse the great contributions of modern science to human welfare, or even to mention the major branches of technology which have profited from advances in fundamental theoretical and experimental research during the past one hundred years. It is sufficient to note that applied science has transformed the face of the earth, and has brought into being our contemporary Western civilization. Although Francis Bacon lived before modern science achieved its most spectacular victories, he was in this regard a true prophet of things to come. His conception of the nature of scientific method is now generally recognized to be woefully inadequate. But he sensed correctly the power over things that was latent in the "new science" of the seventeenth century; and he expressed a fundamental aspiration of his own and of later generations when he declared that "the true and lawful goal of the sciences is none other than this: that human life be endowed with new discoveries and powers." Since the technological fruits of systematic inquiry can be appreciated even by those with no scientific training, it is this aspect of science which is most prominent in the popular literature and teaching of science. Indeed, the practically useful fruits of fundamental research have in recent years been frequently stressed, even by the researchers themselves, in an attempt to justify the cost to those who ultimately shoulder it by heavy governmental subsidy of apparently otiose scientific inquiry.

I do not share the outlook of those who minimize the value of the material goods which science as an improved and expanding technology contributes toward the enhancement of human life, and who attribute the so-called "materialism" of our age to the growth of scientific technology. I think, nevertheless, that science viewed as a set of practical controls over nature has been overemphasized in recent years, to the neglect of other of its aspects. It is certainly not the case that the achievement of immediately useful ends is the sole, or even the central, motive which actuates scientific inquiry. A false picture is created of the history of science and of the complex goals of systematic inquiry when that motive is made focal. Moreover, this emphasis tends to portray science as a miscellany of surprising wonders and gadgets, and to create an image of the scientist as a miracle-worker with a nostrum for every physical and psychic ill, whose opinions, like those of successful businessmen and military leaders, are to be taken as authoritative even on matters about which they have demonstrable incompetence.

Nor can we afford to be complaisant about the widespread though mistaken tendency to make fundamental research responsible for the social consequences of applied science—an imputation of responsibility which appears plausible when science and technology are identified, and which is often used to justify attempts at restricting the freedom to inquire when those consequences are admittedly undesirable.

2

In any event, science is more than a set of practically useful technologies. Science has another facet, consisting of a body of conclusions, stated in propositional form and often general in content, which are asserted as the warranted findings of special inquiries. According to an ancient formula, the aim of theoretical science is to "save the phenomena," that is, to exhibit the events and processes occurring in the world as instances of general principles which formulate invariable patterns of relations between things. Science as a body of universal law or theories does "save the phenomena" in this sense, and makes the world intelligible. In achieving this objective, science satisfies the craving to know and to understand, perhaps the most powerful impulse leading men to engage in systematic inquiry. In achieving this objective, science is a primary force in the history of man for undermining superstitious beliefs and practices, and for dissolving fears which thrive on ignorance. Science as theoretical understanding is the creator of the intellectual basis for a humane and liberal civilization: it supplies the indispensable ground for evaluating inherited customs and for assessing traditional principles by which men habitually order their daily lives.

It was once taken for granted that no proposition is genuinely scientific unless it is demonstrable from first principles. It was assumed, moreover, that the hallmark of scientific truth is truth capable of being known with complete certainty and capable of being apprehended as absolutely necessary. This conception of the nature of science took geometry as its model. A little reflection shows, however, that not everything can be demonstrated, and that some premises must be accepted without demonstration. Accordingly, so classical rationalism maintained, if the ideals of certainty and necessity are to be realized by a science, the basic premises or first principles of the science must be grasped in an immediate originative act of the intellect as necessarily and self-evidently true.

The conception of scientific knowledge just outlined was once plausible. Subsequent developments have shown, however, that the body of scientific conclusions is not fixed once and for all—not merely because additions to it are made by the progress of research, but also because modifications in previously held views are frequently required. This point needs no elaboration in the light of the recent history of physics, which has witnessed fundamental revisions of theoretical assumptions that were once regarded as indubitable. Nevertheless, to those whose notions of science have been formed within the intellectual framework of classical rationalism, such revisions are often construed as signs of the "bankruptcy" of modern science. A wholesale skepticism concerning the possibility of obtaining genuine knowledge by way of science has in consequence become widespread. In their quest for unshakable certainty, many have turned elsewhere for a type of knowledge which the sciences do not provide, but which other modes of alleged knowing profess to supply.

There is another feature of modern scientific theory which calls for comment. The abstract formulations required to express the comprehensive conclusions of modern inquiry often make the relevance of those conclusions to the facts of gross experience far from evident;

and the scientific objects postulated by current scientific theory seem to be not only utterly remote from the familiar things of daily life, but also to constitute an order of reality radically different from the reality of customary objects encountered in common experience. Readers of the late Sir Arthur Eddington's *The Nature of the Physical World* will doubtlessly recall the contrasting picture he drew between the familiar piece of furniture called a table and the "scientific" table. The former was described by him as extended, permanent, substantial, capable of supporting heavy objects, and possessing a great variety of esthetic qualities. The latter was characterized as being mostly emptiness, a congeries of rapidly moving and invisible electric charges separated by enormous relative distances, insubstantial, and possessing neither colors nor textures.

Given such an account of what are allegedly two tables, it is difficult to avoid raising the question of which one is the "reality" and which the "appearance." Many scientists, including Eddington, have repeatedly raised it, in connection with physics as well as with other parts of scientific theory. Some have argued that our ordinary, common-sense encounters with the world yield only illusions, and that the objects discovered and formulated by theoretical physics are the exclusive reality. Others have maintained the contrary thesis, and have insisted that scientific theories are simply convenient fictions possessing no cognitive value. But whatever the answer proposed, the consequences of introducing such a distinction between the real and the apparent have been intellectually disastrous. For the supposition that such a distinction is relevant has, in effect, alienated man from the rest of nature, and has reinforced a skepticism concerning the possibility of acquiring genuine knowledge about the world by way of the sciences. Despite the increased practical control over things which the advance of science yields, when that distinction is drawn the world is made progressively more mysterious and strange by the development of theoretical analysis. It is not surprising, therefore, that instead of becoming a source of enlightenment, the theoretical conclusions of science have frequently become means for strengthening the hold of ancient dogmas and for sanctioning the adoption of antirational philosophies of life.

The point of these remarks is that the conclusions of science are not fully intelligible without a clear sense of the logical relations in which they stand to each other as well as to the facts of gross experience, and without a clear conception of the intellectual method by which conclusions are certified as valid. A philosophy of science that undertakes a critique of the abstractions of science and that analyzes the nature of this method, therefore, seems to me indispensable, both for an adequate appreciation of the substantive content of science, as well as for a sober appraisal of the import of that content for the principles by which men order their daily lives.

3

This brings me to the third facet which science exhibits, namely, its method of inquiry. I have already noted that every special finding of science, whether it be general or particular in form, may require reconsideration because of fresh evidence that seems to challenge it. No question raised for study is ever settled beyond the possibility of its being reopened, for no proposed answer is indubitable. Indeed, when one surveys the history of science, the impermanence of nearly all of its comprehensive theories is most impressive. As I see it, it is the method of science that is relatively stable and permanent, rather than the answers to questions accepted at various times. It is certainly the case that the alleged certainty of scientific knowledge derives from the intellectual method by which the findings of inquiry are warranted.

To prevent misunderstanding, however, I must state explicitly that in my present usage, the word "method" is not synonymous with the word "technique." The technique of measuring wave lengths with a spectroscope is patently different from the technique of measuring the speed of a nerve impulse, and both are different from the techniques employed in assessing the date of composition of some historical document. In general, techniques differ with subject matter and may be altered with advances in technology. But the general logic of sifting and evaluating evidence, the principles employed in judging the adequacy of proposed explanations, the canons implicit in making responsible decisions between alternative hypotheses, are the same in all inquiries. These rules of *method* may be modified as science develops. Nevertheless, changes in them are usually infrequent and relatively minor, so that there is a recognizable continuity between the logic of proof used by scientists living at different times and places. It is these general principles of evaluating evidence that I have in mind when I speak of scientific method. Since in my view of it the method of science is the most distinctive and permanent feature of the scientific enterprise, I propose to offer a brief account of some elements in that method.

It is appropriate to begin this account with the reminder that science is a social institution, and that professional scientists are members of a self-governing intellectual community, dedicated to the practice of modes of inquiry which must meet standards evolved in a continual process of mutual criticism. It is often supposed that the objectivity and validity of conclusions reached in the sciences are the products of individual intellectual insights, and of individual resolutions to believe only what is transparently true. Thus, the seventeenth-century philosopher and mathematician Descartes proposed a set of rules for the conduct of the understanding, adherence to which would guarantee that the mind would not fall into error. He stipulated that nothing is to be accepted as genuine knowledge which is not absolutely clear and distinct, or which is in the least manner doubtful. Descartes thereby assumed that scientific truth is simply the fruit of the operation of individual minds, each relying exclusively and separately on its own power to discover what is indubitable. Descartes' rules, however, are counsels of perfection, and have little effective value. Men are usually unaware of all the tacit assumptions involved in what they think is indubitable, and they frequently suppose themselves to be making no intellectual commitments of any kind, although in fact they are tacitly subscribing to much that is false. Accordingly, while pious resolutions on the part of scientists to be critical of their own assumptions may have a certain value, the objectivity of science is not primarily a consequence of such resolutions. On the contrary, assured knowledge in science is the product of a community of thinkers, each of whom is required by the tacit traditions of his group to be unsparing in his criticisms and evaluations of the cognitive claims presented to him.

It is through the operation of continual and independent criticism that scientific truth is achieved. The tradition of science is a tradition of toleration for new ideas, but a toleration qualified by a sturdy skepticism toward any notion which has not been subjected to observational and logical tests of validity—tests which are intended to be as severe as human ingenuity can devise. No one scientist engaged in this process of criticism is infallible, and each one will have his own peculiar intellectual or emotional bias. But the biases are rarely the same; and ideas which can survive the cross fire of the varied critical commentary that a large number of independently acting minds supply stand a better chance of being sound than conceptions which are alleged to be valid simply because they appear to be self-evident to some individual thinker.

It follows that the source or origin of scientific ideas is entirely irrelevant in evaluating their empirical validity. The adequacy of scientific hypotheses is established ultimately by reference to matters capable of repeated public observation; and however great may be the honor and reverence scientists pay to the creators of scientific ideas and to the discoverers of new phenomena, it is not the individual authority of the great innovators which warrants the acceptance of their contributions. Moreover, there is nothing too sacred or too lowly, too unusual or too commonplace to be exempt from the critical scrutiny of science; and in this regard scientific inquiry respects neither persons, places, traditions, nor mysteries. According to St. Augustine, the question of what God was doing before Creation deserves something better than the mocking answer that He was preparing hell for those who pry into mysteries. But however this may be, if that answer is not just a joke it illustrates an attitude that is radically incompatible with the tradition and temper of mind of modern science.

Nevertheless, despite its deliberate cultivation of a tough-minded critical spirit, the scientific attitude is not a wholesale skepticism concerning the possibility of genuine knowledge. On the contrary, the endless criticism in which the scientific community is engaged is performed on the assumption that reliable knowledge *can* be achieved by way of just such a critical process. Indeed, reliable knowledge is in fact the product of science; and many of its conclusions, even when universal in scope, have stood up under centuries of severe testing. Although no conclusions of science are beyond the possibility of doubt, and all are corrigible in the light of future evidence, not all proposed conclusions of inquiry have to be abandoned as false. The fact that we *may* be in error, and that we must be ever on the lookout for discrepancies between our beliefs and what actually occurs, does not mean that we are always mistaken and that we know nothing. On the contrary, though the conclusion of any given scientific inquiry is in principle always subject to correction, in actual fact the content of science does not consist chiefly of short-lived opinions in constant flux. Numerous examples could be mentioned of scientific findings that have long endured in the face of continued study, but a single one must suffice. Galileo's conclusions concerning the strength of beams have been made more precise and extended by subsequent investigations. However, his explanation of why there are no giant creatures with limbs and trunks, whose relative dimensions are like those of human beings but whose absolute dimensions are vastly greater, continues to be regarded as substantially correct.

But I must turn to other phases of scientific method. According to a popular view, sometimes endorsed by distinguished scientists, science starts an inquiry by collecting facts, and then passes the data through some sort of logical sieve which yields a uniquely determined formulation of a regularity between phenomena. This is a seriously misleading account of what actually takes place. For inquiry begins with a *problem*, provoked by some practical or theoretical difficulty; and in general it is not easy to know just what facts one ought to gather to resolve the problem, or whether a purported fact really is a fact. Just what facts ought to be collected in studying the causes of rheumatoid arthritis? Is there really such a phenomenon as extrasensory perception, as many claim? The number of presumptive facts is legion, and we cannot note all of them. The scientist must therefore be selective, and concern himself only with those which are relevant. In consequence, he must adopt a preliminary hypothesis or guess as to how his specific problem may be resolved; and he must employ that hypothesis to suggest what he hopes are the relevant facts. In short, a responsibly conducted inquiry is guided by ideas that must be supplied by the investigator. Observation and experiment serve to test or control the adequacy of those ideas for the problem at hand; but observation and experiment do not provide the conceptions without which inquiry is aimless and blind.

It is a mistake to suppose, therefore, that the conclusions of scientific inquiry are uniquely determined by the facts of observation or experiment. In particular, it is of central importance to recognize that there is no logical route leading from data of observation to the explanations eventually adopted for them. As Einstein has repeatedly observed, the comprehensive theories of modern physics are "free creations of the mind," and require for their invention feats of imagination quite analogous to creative effort in the arts. Consider, for example, the motions of the planets, which have been observed for thousands of years. Neither the hypothesis that the planets move around the earth, as Ptolemy supposed, nor the hypothesis that they revolve around the sun, as Copernicus proposed, can be read off from what is directly observed.

Observations must be interpreted, and the interpretation involves the introduction of general ideas which are supplied by the creative imagination of the scientist. The common failure to recognize the role of this creative element in science is at the bottom of the charge, so frequently leveled against modern physical theories, that they are "unintelligible" because they are incompatible with "common sense"—that is, with ideas that are the deposit of a prior stage of scientific development, but which have become so familiar that they have assumed the status of necessities of thought. It is this failure also which is a partial source of the prevailing skepticism concerning science as an avenue to genuine knowledge. This skepticism has its roots in the naive assumption that certainty in knowledge is a matter of immediate apprehension or direct vision, so that the lack of self-evidence in the explanatory principles of science is taken as a mark of their fictional character.

Nevertheless, though scientific theories cannot be read off from the observed data, science is neither poetry nor pure mathematics, and the validity of a theory is established only through verifying it by experiment or observation. But a theory that is worth anything cannot be tested directly. Its consequences must first be explored with rigorous logic and eventually confronted with the outcome of controlled observations. Many students have argued, under the influence of the writings of Professor Bridgman, that all the notions which enter into a scientific statement, if the statement is to be meaningful, must be given explicit definitions in terms of what can be directly observed; and it is often maintained, especially by those concerned with less developed disciplines like psychology, that a proposed explanation of phenomena is to be rejected as meaningless if it contains terms for which operational definitions are not provided. These are, however, unduly severe requirements for scientific constructions, and little reflection is needed to realize that some of the most successful theories in the natural sciences do not satisfy them. For example, neither the term "electron" in current physical theory, nor the term "gene" in genetics are defined by way of an explicit operational procedure. It is therefore clearly not essential that the theoretical notions of a science should refer directly to matters identifiable in gross experience. Indeed, the explanatory power of theories is in general directly proportional to the remoteness of their key concepts from things capable of direct observation. What is essential, however, is that at some point in the deductive elaboration of a theory, suitable links be established between some of its concepts and experimentally identifiable properties of macroscopic objects.

It is for reasons just noted that deduction in general, and mathematics in particular, play such an important role in modern science. For mathematics is the great art of drawing rigorous conclusions from any set of clearly formulated postulates; and the significance of an explanatory theory can be determined only through the discovery, by way of deductive inference, of just what its principles do imply. It is an illusion to hold, however, as even distinguished physicists sometimes have, that the mathematical language in which current physical theories are formulated shows that the structure of things is essentially mathematical and that therefore the world must have been created by a divine mathematician. The structure of processes in *any* world, whether actual or conceivable, could always be formulated in an appropriate mathematical notation. But the *postulates* of a theory about factual subject matter are not *mathematical* truths, any more than is the statement that I possess two hands a mathematical truth. The postulates of a theory can be certified as valid, not by establishing their mathematical necessity, which is clearly impossible, but by showing that their logical consequences are in some measure in agreement with data of observation.

A detailed account of the logic of testing hypotheses cannot be given here; but brief mention must be made of the notion of a *controlled inquiry*—perhaps the most important single element in that logic. A simple example must suffice to indicate what characterizes such inquiries. The belief, widely held at one time, that cold salt water baths were beneficial to patients suffering from high fevers, seems to have been based on repeated observations of some improvement in the condition of patients subsequent to such treatment. However, irrespective of whether or not the belief is sound—in point of fact it is not—the evidence on which it was based is insufficient to establish it. It apparently did not occur to those who accepted the belief on that evidence to ask whether patients not given such treatment might show a similar improvement. In short, the belief was not the product of a controlled inquiry—that is, the course of the disease in patients receiving the treatment was not compared with its course in a "control" group of patients who did not receive it, so that there was no rational basis for deciding whether the treatment made any difference.

More generally, an inquiry is a controlled one only if, by instituting some kind of *eliminative* procedure, the *differential* effects of a factor that is assumed to be relevant to the occurrence of a given phenomenon can be ascertained. Such an eliminative procedure is sometimes, but not necessarily, an overtly experimental one, since in many domains of inquiry (as on the weather or of human affairs) overt experiments are in general not feasible; and in consequence, quite complicated and subtle analytical tools must frequently be used to extract from the available evidence the information needed for a rational decision on the merits of a hypothesis. But, in one form or another, the notion of control is an essential ingredient in the logic of scientific method—for by and large, the reliability of scientific conclusions is a function of the multiplicity and the rigor of the controls to which such conclusions have been subjected.

It is frequently asserted that because of the introduction of quantitative distinctions and measurement, the natural sciences in effect ignore the diverse qualities of common experience and fail to assign to the latter any place among the constituents of the world. How ill-founded such dicta are is readily revealed by an analysis of the nature of measurement. Instead of ignoring or denying the existence of qualitative distinctions, techniques of measurement both presuppose qualitative differences as well as make it possible to recognize more subtle qualitative distinctions.

A simple example will make this point clear. Human beings are able to distinguish between a certain number of degrees of warmth, and terms like "hot," "very warm," "warm," "lukewarm," "cool," and "cold" correspond to these familiar differences. But these differences were not ignored or denied when the thermometer was invented in the seventeenth century. The thermometer was developed when variations in experienced differences of warmth of many substances were found to be connected with changes in the relative volumes of those bodies. In consequence, variations in volume could be taken to represent changes in the physical state of a substance, which in some cases correspond to the felt differences in their degree of warmth. It is possible, however, to recognize finer differences in variations of volume than in directly experienced changes in degrees of warmth; and there are extremes of heat and cold beyond which human beings are unable to discriminate any further, although volume changes beyond these limits can still be distinguished. It is therefore evident that the use of the thermometric scale does not involve ignoring qualitative differences; that scale merely orders qualities in an unambiguous manner, and provides a way of noting differences among them which would otherwise escape attention.

When the conclusions of the sciences are thus analyzed in the light of the logic of scientific procedure, it also becomes manifest that those conclusions do not assert that any of the objects or qualities of common experience are "unreal," and that when a theory succeeds in explaining the events around us, it does not explain them away. The occurrence of rainbows, for example, is explained in terms of the fundamental ideas of geometric optics. But it would be fatuous and self-stultifying to maintain as a consequence of this explanation that there are no such things as rainbows, or that only optical rays are "real" while the colors observed in definite patterns in the sky are "unreal." Indeed, no questions as to what is real or unreal can be significantly raised in such contexts. What the explanation achieves is the incorporation of the special phenomenon under investigation into a wider net of relations, so that the conditions for the occurrence of the phenomenon are shown to be similar to thc conditions of occurrence of a large class of other phenomena as well.

Again, when the macroscopic properties of a gas are explained in terms of the molecular theory of matter, so that in effect the laws of gases are "reduced" to theorems in the theory of statistical mechanics, this "reduction" cannot intelligibly to construed to deny that gases have temperatures, pressures, diffusion rates, and other properties. The reduction consists in exhibiting the regularities established for gases as special cases of more comprehensive uniformities, in terms of which many other special regularities can be shown to be systematicaly related. In short, the progress of science does not consist in the denial that the familiar faces of things are genuine phases or parts of existence. The progress of science consists in discovering connections and relations of order between things encountered in experience, thereby yielding intellectual, if not always practical, mastery over the events and processes of nature.

4

According to the well-worn cliché, we are living in an age of science. But the great hope, widespread before the first World War, that science was a force which would bring into being a humane social order and at the same time free men's minds from ancient superstitions, has noticeably weakened. Many men have turned away from the conception of reason embodied in the scientific process, and have come to place their faith in various forms of irrational philosophies of life as guides to salvation. There is no simple explanation for these changes in outlook, and I can mention only a few of the factors that have contributed to bringing them about. The cataclysmic wars and violent social upheavals of the past four decades have also weakened the appeal which the less dramatic ways of a cautious scientific intelligence once had. Those in seats of power, moreover, have not always been willing to use the knowledge and skills available to improve the estate of mankind throughout the world; and men who spend their lives in abject poverty and physical misery cannot be expected to wait indefinitely for the realization of dreams that are never implemented. Again, many of those who had pinned their hopes for mankind on the steady progress of scientific enlightenment undoubtedly expected too much too soon. In looking forward to the early embodiment of heaven on earth they were bound to be severely disappointed. Furthermore, the technological use of science in warfare and the pseudoscientific conceptions with which various social movements have dressed up their policies have persuaded many that science *per se* is not an unalloyed good but a potent source of evil. Scientists in their capacity as seekers after theoretical understanding have sometimes been held responsible for the barbaric use to which modern society has put their discoveries.

There is one further reason for the decline of the great hope in science which seems important to me. This is the fact that, in the main, science has been taught, both on popular and advanced levels, either as a set of technologies or as a body of conclusions, without exhibiting them as products of the intellectual method which is the lifeblood of modern science and a supreme agent of liberal civilization. I am convinced that the general failure to present scientific conclusions in the light of the logic of scientific procedure is the source of much confusion and a cause of much contemporary obscurantism. For I am persuaded that a large potential audience is eager to learn just what is the import of recent theoretical researches for a vision of human life. The exclusive emphasis so often placed on applied science does not satisfy this interest. Man is among other things a speculative animal. If he does not obtain from scientists an adequate conception of the scheme of things entire and of man's place in it, he will turn to other sources, whatever be their credentials, for the materials for constructing such a vision. Scientists have done little to meet this legitimate need, in considerable measure because they are not trained to do so. Most scientists acquire the temper of mind and the intellectual method essential for competent research not by explicit formal teaching but from example and by acquiring suitable habits of workmanship. Few scientists are vitally concerned with methodological issues, or with the logic of their discipline. When they do venture to discuss general ideas of the logic of procedure connected with their professional work, as they often do on ceremonial occasions, most of them often echo the uncriticized but questionable philosophies they learned at their mother's knee.

5

I do not want to leave these observations entirely unsupported by evidence, and will therefore cite some misleading interpretations of natural science disseminated by leading scientists as well as philosophers.

Perhaps the most serious misinterpretation consists in the widely repeated claim that modern science is no longer "mechanistic," and that both the physical theory of relativity as well as quantum theory indicate the advent of a new scientific method which acknowledges the spiritual character of reality. Now it is indeed the case that the science of *mechanics*, which was once held to be the universal science adequate for explaining all processes in nature, has lost its preeminence; and no physicist today believes it is possible to understand the great variety of electrical and radiational phenomena in terms of the fundamental principles of Newtonian theory. But it is one thing to say that the science of mechanics is no longer regarded as the fundamental science, and quite a different thing to claim that science is in no sense any longer "mechanistic"—that is, that it no longer seeks to discover the mechanisms and the conditions involved in the occurrence of events and processes. Surely the latter is a grotesque claim, belied by every paper published in scientific journals. And just as surely, the recent recognition that mass is convertible under certain circumstances into energy, so that while the principle of conservation of energy has universal validity the principle of conservation of mass has only a qualified scope, does not signify that what is loosely called "matter" has disappeared from the universe or that some sort of "mental" stuff has taken its place. It is absurd to suppose, moreover, that whenever some theory in a science is discovered to be inadequate for a given domain of phenomena and is replaced by some other theory, a change in intellectual method of conducting inquiry is necessarily involved. Even a casual comparison of the logic of proof which dominated nineteenth-century physics and the logic which current science employs makes abundantly clear that, despite alterations in the substantive concepts employed in the respective theories, there has been no comparable change in the modes of evaluating evidence.

An analogous misconception has become current concerning the alleged "breakdown" of the principle of causality, because of the so-called "indeterministic" character of quantum theory. In particular, the famous Heisenberg Uncertainty Relations, derivable within that theory, have been hailed as establishing the "reality" of free will not only for human beings but also for electrons and other submicroscopic particles of physics. Now there is no doubt that quantum mechanics possesses features that sharply differentiate this theory in its internal structure from the theories of classical physics. For in one standard interpretation, quantum mechanics, unlike the theories which preceded it, is an essentially statistical theory, so that within the framework of this interpretation the theory establishes fixed connection only between *statistical* properties of aggregates, not between properties of the *individuals* which constitute those aggregates. But it is nonetheless doubtful whether this fact can be taken as evidence for the collapse of the causal principle, especially if the latter is construed (as it in fact must be, if it is to have a universal scope even within classical physics) as a methodological or formal principle rather than as a substantive one.

Within quantum mechanics itself, as has just been noted, there are determinate and invariable relations between certain complex properties of things, even though no such relations are asserted to hold between other properties. Moreover, waving this point, quantum mechanics is so related to theories about macroscopic objects that in the limiting case, as the physical systems to which quantum mechanics is applied increase in their dimensions, quantum mechanics coincides with the theories of macrophysics. Accordingly, even if the submicroscopic particles do not satisfy what would be recognized as deterministic causal laws, it does not follow that *macroscopic* objects do not satisfy such laws. The supposition that quantum mechanics proves the general breakdown of causality is therefore a caricature of the actual situation. In any case, the ordering of our practical affairs, as well as the analysis of the extensive range of phenomena for which classical physics is still undoubtedly valid, can be successfully conducted only if we pay due heed to the fixed dependencies known to hold between events and the determinate conditions of their occurrence.

Moreover, an elementary confusion is committed when human freedom is construed in terms of the alleged "indeterminacy" of electrons and other physical particles. Human freedom is constituted by the capacity of human individuals to choose and act, when no external compulsions or constraints are present, in the ight of their desires and the evidence available to them. It is a capacity which is fully compatible with either classical or quantum physics. Whether a human being possesses freedom in this sense can in general be ascertained by examining the circumstances of his overt behavior. The introduction into this context of considerations drawn from quantum theory is entirely irrelevant. Thus, before his imprisonment, Socrates was free to converse with the Athenian youth in the market place of ancient Athens; but after his condemnation by the Athenian court he was not free to do so. Neither the replacement of classical mechanics by quantum theory, nor the possibility that the latter may some day be succeeded by a strictly causal theory of radiation phenomena, in any way alters these facts.

There have also been persistent attempts in recent years to reintroduce organismic and teleological interpretations of all natural processes, inorganic as well as organic. Some of these attempts are based on representing current physical theory in terms of conveniently visualizable models, which are then described in anthropomorphic and highly misleading language. The old Bohr theory of atomic structure, for example, is usually presented as postulating a number of electrons revolving on definite orbits around a nucleus. According to this model, an electron may jump from one orbit to another, and so may either emit or absorb quanta of energy, provided that the second orbit is not already occupied by another electron. This supposition is then sometimes described as implying that, in leaving its original orbit, an electron must "know" in advance where its jump can terminate. Accordingly, atomic processes are said to be intelligible only in terms of teleological notions, essentially similar to those involved in accounts of human behavior.

More generally, it is frequently maintained that the motions and changes in nature are not "blind" and "purposeless," but that the universe is an "organism" whose parts are purposively related and exhibit modes of action quite analogous to those found in humans. Indeed, distinguished scientists and philosophers have sought to base their conception of a desirable society upon the assumed facts of subatomic structure; and some of them have even claimed that the thermodynamics of physical systems is evidence for the validity of Christian morals.

In my judgment, these various attempts to reinstate teleological notions into physical science are guilty of debasing and misusing language, in the interest of a spurious consolation which the use of such notions supposedly brings. To say that an electron must "know" into which orbit it will jump, is to use the word "know" in a sense radically different from its familiar connotation, with the misleading suggestion that there is some analogy, otherwise unspecified, between its familiar and its extraordinary meaning. Similarly, to declare that physical systems in general are purposively organized either involves the use of the word "purposive" in a new but undefined sense, or it wipes out the manifest differences between animate and inanimate systems. Nothing but confusion is achieved when purely physical processes are formulated in language which is commonly used to characterize traits that are distinctive of human beings and other animals. The glaring differences between purely physical change and purposive human behavior are neither reduced nor illuminated by such descriptions. Modern physics has indeed discovered in the domain of the subatomic hitherto unfamiliar types of structure and modes of organization; but when subatomic processes are characterized as forms of teleological behavior, language is being employed in a dangerously irresponsible manner. As the eighteenth-century Bishop Butler wisely remarked, "Things are what they are, and their consequences will be what they will be; why then should we desire to be deceived?"

I suspect that there is a will to illusion in all of us, and that we sometimes desire to be deceived. But I doubt whether the misconceptions I mentioned are, in general, the products of such a desire. On the contrary, I am of the opinion that most of the misconceptions concerning the content and import of science are in large measure the result of the pervasive failure to present science in a manner which makes explicit the logic of scientific method, the complex but flexible relations in which scientific abstractions stand to matters of gross experience, and the historical development of our efforts to understand the world around us. To make these things explicit is one task of the philosophy of science, though whether this task is undertaken by the philosophical practicing scientist or the professional philosopher is of little moment.

The philosophy of science aims to effect both a systematic critique of scientific abstractions, and a comprehensive analysis of the components which enter the intellectual method by which reliable knowledge is acquired. It must therefore examine in detail the role of mathematical and other symbolical techniques in the construction of systems of explanation, and so help to identify what is conventional or definitional in the corpus of scientific statements, as distinct from what must be warranted by reference to empirical fact. It is of utmost importance to avoid the still widely prevalent confounding of statements which are expressions of laws of nature, and statements which are logically necessary but are without empirical content. For example, a moment's reflection suffices to show that the statement "If no mammals are invertebrates, then no invertebrates are mammals" is not an expression of a biological law, since the statement can be certified as valid by means of purely logical operations. Nevertheless, comparable though more complex statements, expressing truths of logic but void of empirical import, have often been paraded as important conclusions of theoretical research in both the natural and social sciences.

A philosophic analysis of science also focuses attention on the various functions which theoretical constructions possess in inquiry, and thereby provides guards against the tempting fallacy of imputing meanings to statements which are irrelevant to the role those statements play in given contexts of usage. The philosophy of science also undertakes to evaluate the relative merits of alternate inductive policies, and to articulate the rational basis for the acceptance or rejection of proposed hypotheses about suspected regularities in nature. Moreover, it may attempt some form of synthesis of the special findings of scientific inquiry, either by constructing an inclusive account of the universe, or by formulating a set of basic categories into which everything that exists will supposedly fit—although in my opinion such attempts have been generally of dubious worth.

Finally, the philosophy of science may occupy itself with the relations of science and the rest of society. It may then examine the social determinants of scientific progress, the technological and ideational impact of scientific changes upon institutions and moral ideals, and the still unrealized possibilities for human betterment that are implicit in the funded scientific knowledge of mankind.

In any event, there is an imperative need for viewing science not simply as a set of technologies or as a body of conclusions, but as a continuing process of inquiry whose fruits are the products of a remarkable intellectual method. The need is imperative, if the age-old and socially costly conflict between the sciences and the humanities is ever to be overcome.

Interest in the humanities is sometimes associated with a snobbish traditionalism, a condescending attitude toward whatever is modern, and an ill-concealed contempt for rigorous, critical thought. There are in fact professed humanists who are lost in admiration for the sewage systems of the ancient Romans, but dismiss as beneath notice modern works of public hygiene. There are self-proclaimed humanists who are profoundly moved by the account in Herodotus of the heroism of the Greeks at Thermopylae, but regard the courageous uprising in 1944 of Warsaw Jews in which they were decimated by their Nazi oppressors as just another minor historical event that is hardly worth notice. There are humanists today who judge science to be an essentially menial occupation, and who see nothing but trivialities in current attempts at logical, cogent analyses of the rationale of moral preference.

Nevertheless, those who exhibit these attitudes are not, in my judgment, the best representatives of the values for which "humanism" has long been a label. The passionate interest in whatever is human, whether it be a great work of imagination or a humble manifestation of man's feeling for his kind, as exhibited by such men as Leonardo, Erasmus, Spinoza, John Stuart Mill, or Santayana in our own day, is marked by no such narrowness of spirit or such intolerance of what is fresh and nascent. On the contrary, what is distinctive of humanism at its best is a judicious temper and balanced judgment, a disciplined readiness to profit from fresh experience, a broad angle of vision and avoidance of fanaticism, and a healthy scepticism toward what has not withstood the test of experience.

But if these are the traits essential for a genuine humanism, there can be no conflict between the scientific and the humanistic attitudes. For in their ideal form, a life devoted to science and a life devoted to the humanities exhibit a common temper of mind. There undoubtedly is an acute antagonism between many who pursue science and many who pursue the humanities—an incongruence which is in part the result of the extraordinary but inevitable specialization that has become so necessary for the successful pursuit of modern scientific inquiry, in part a fruit of the aloofness and analogous concentration that characterizes humanistic studies. But I venture the hope that this opposition can in considerable measure be mitigated, from one direction, at any rate, by exhibiting science as a form of activity which demands from those engaged in it the same kind of creative sensitivity, discriminating toleration, and reasoned understanding which humanistic studies at their best also require. Nor is it possible to dismiss the cardinal fact that the sciences, no less than the humanities, have much to contribute on the great issues which concern men as men. For the sciences provide the intellectual substance out of which we can construct an authoritative vision of nature and of man's place in it. Such a vision, based on the clearly and soberly understood findings of the sciences, may offer little comfort to those who are content with nothing less than a universe that is mindful of human needs. Nevertheless, such a vision has been in the past, and may once more become in the future, a source of spiritual strength and stability, and a dependable guide in the pursuit of reasonable human ends.

It must be both admitted and emphasized, however, that science does not exhaust the modes of experiencing the world. The primary aim of science is *knowledge;* and however desirable and precious this fruit of science may be, it clearly is not and cannot be a substitute for other things which may be equally precious, and which must be sought for in other ways. On the other hand, although what science has to offer is understanding, and although that understanding may be instrumental for gaining something else, knowledge is a final value, self-justifying because it enhances human power and feeds an enduring enjoyment. Science does satisfy what is perhaps the most distinctive human desire, the desire to know; and no one who is genuinely devoted to the humanities can ignore the dimension of experience to which science as the quest for knowledge is relevant. It satisfies that desire by dissolving as far as it can our romantic illusions and our provincialisms through the operation of a social process of indefatigable criticism. It is this critical spirit which is the special glory of modern science. There are no reasonable alternatives to it for arriving at responsibly supported conclusions as to where we stand in the scheme of things and what our destinies are.

2/Theory and Observation

In his well-known Herbert Spencer Lecture "On the Method of Theoretical Physics," delivered in 1933, Einstein took as the starting point of his discussion what he characterized as "the eternal antithesis" between the rational and the empirical components of physical knowledge, an antithesis he reaffirmed in his subsequent reflections on scientific method. He pointed out that a physical theory is an ordered system of certain fundamental general postulates or assumed laws, stated in terms of various basic concepts dealing with the matters under study, and he stressed the indispensable role of logical deduction in making explicit the numerous detailed consequences of those postulates. He also underscored his conviction that while such concepts and assumptions may be, and perhaps usually are, *suggested* by the things investigated in physics, they are not derivable from those things either by abstraction or by some other logical process, but are "free inventions of the human mind." However, Einstein went on to say, the imaginative creation of ideas and "purely logical thinking cannot yield us any knowledge of the empirical world; all knowledge of reality starts from experience and ends with it. Propositions arrived at by purely logical means are completely empty as regards reality." If a physical theory is to count as a valid account of the actual world, and to be more than the formulation of logical possibilities, the conclusions deduced from its postulates "must correspond" with what experience reveals. "Because Galileo saw this," Einstein continued, "and particularly because he drummed it into the scientific world, he is the father of modern physics—indeed, of modern science altogether."[1] Accordingly, although on Einstein's view neither the basic concepts nor the fundamental postulates of a theory are uniquely prescribed or determined by observation and experiment, and cannot therefore be simply "read off" from what is encountered in sensory experience, the *validity* of a theory as an account of the actual world depends upon, and is controlled by, what is disclosed to observation and experiment.

In emphasizing the decisive role of these latter in the *testing* of physical theories, Einstein was stating a view that has been commonplace for several centuries in discussions of scientific method. Even his claim that the central ideas of modern physical theory are neither abstracted from, nor explicitly definable with finitistic means in terms of, unquestionably manifest traits of the phenomena studied in the science is at present not a seriously mooted issue among philosophers—though undoubtedly the claim contradicts beliefs associated with certain historical forms of empirical philosophies of knowledge, and continues to be included in widespread popular notions about the nature of scientific inquiry. In any event, there are no current analyses of the logic of inquiry in the natural sciences which formally deny the cardinal role of what Einstein called "experience" (i.e., controlled observation and experiment) in assessing the cognitive worth of a scientific theory.

However, despite the absence of such formal denial, the authority of experience in evaluating the adequacy of theoretical assumptions has nonetheless been recently challenged. For the view that a proposed theory in the natural sciences must be tested against the findings of experience appears to rest on the assumption that a clear distinction can be drawn between, on the one hand, statements which supposedly codify the outcome of observations on the subject matters being explored and, on the other hand, the theories about those matters for which the so-called "observation statements" supposedly provide confirming or disconfirming evidence. Moreover, many philosophers of science who think that such a distinction can be instituted with at least approximate precision also believe that observation and theoretical statements frequently have no nonlogical expressions (or subject matter terms) in common; and they therefore maintain that, if such purely theoretical statements are to be relevant to empirical subject matters, theoretical statements must be related to observation ones by way of so-called "correspondence" rules or laws. But these assumptions, especially the first one, have been the targets of vigorous criticism during the past decade, criticisms which have in fact reopened the question whether, and if so to what extent, theories are subject to the test of experience.

The main burden of these criticisms is that the belief in an "absolute" distinction between observation and theory is untenable. Observation statements, so the criticisms commonly maintain, are not unbiased formulations of supposedly "pure" materials of sensory experience, but involve interpretations placed upon the sensory data, and are therefore significant only by virtue of some theory about the things under study to which the observers are antecedently (though not necessarily permanently) committed. Moreover, theories are ostensibly "free creations" of the scientist, and while their adoption may have definite causal determinants, none of them is logically prescribed by the sensory data. The import of every observation statement is therefore determined by some theory that is accepted by the investigator, so that the adequqcy of a theory cannot be judged in the light of theory-neutral observation statements. Accordingly, if these criticisms are sound, they apparently lead to a far-reaching "relativism of knowledge," to a scepticism concerning the possibility of achieving warranted knowledge of nature that is much more radical than the relativism associated with the views of Karl Mannheim and other sociologists of knowledge. For according to Mannheim, for example, the content as well as the standards for evaluating the validity of a theory in the social sciences (though not in the natural sciences) reflect the social biases and vary with the social perspectives of the individuals who subscribe to the theory. These are, nevertheless, biases which are generated by causal factors operating in certain branches of inquiry, biases which even Mannheim recognized could in principle be neutralized. But the relativity of knowledge implicit in some recent trends in the philosophy of science is of a more fundamental sort. The scepticism it entails concerning the possibility of deciding on objective observational grounds between alternative theories is based on its analysis of the intrinsic structure of human cognitive processes.

But however this may be,this paper is a critique of some of the reasons that have been recently advanced for questioning the authority of even painstaking observation in testing theories, and for supporting a view of science that makes genuine knowledge of nature highly problematic. This sceptical import of recent discussions in the philosophy of science seems antecedently incredible, and indeed I hope to show that the evidence for such a conclusion is far from compelling. It would be idle to pretend, however, that there are no difficulties in drawing a distinction between observational and theoretical statements; and I certainly do not know how to make such a distinction precise. Nevertheless, I do not consider that this distinction is therefore otiose any more than I believe that the fact that no sharp line can be drawn to mark off day from night or living organisms from inanimate systems makes these distinctions empty and useless.

At the same time, I readily admit that the position for which I shall be arguing is essentially a middle-of-the-road one, and is certainly not novel. For on the one hand, I agree with the view repeatedly advanced in the history of thought by critics of sensationalistic empiricism—though proclaimed as a novel idea by some recent commentators on the logic of science—that the sense and use of predicates employed in the sciences, including those employed to report allegedly observed matters, is determined by the general laws and rules into which those predicates enter. In consequence, the content of an observation statement is not identifiable with or exhausted by what is "directly" encountered in any given sensory experience, so that every such statement is corrigible "in principle" and may be revised (and perhaps even totally rejected) in the light of further observation and reflection.

On the other hand, I also think that many observation statements acknowledged in controlled inquiry as well as in the normal affairs of life do not *in fact* require to be corrected; that the various laws which determine in whole or in part the content of such statements are often so well supported by evidence that they are beyond reasonable doubt; and that the content of observation statements is not in actual fact determined by the *totality* of laws and rules of application belonging to the corpus of assumptions of a science at a given time. But if this is so, the meaning and validity of observation statements are in general not determined by the very theory that those statements are intended to test; and accordingly, such statements can be used without a stultifying circularity to assess the factual adequacy of the theory. However, these claims need to be supported by argument, a task to which I now turn.

1

The words "theory" and "observation" as well as their various derivatives are notoriously ambiguous as well as vague. Indeed, although the expressions "theoretical statement" and "observational statement" are often associated with ostensibly contrasting senses, there are also contexts in which they are used with overlapping denotations. For example, the adjective "theoretical" is sometimes employed to refer to statements about microentities or microprocesses, such as statements about the motions of electrons, on the assumption that these matters cannot be perceived (in the sense in which an experimental psychologist understands the word "perception"), while the term "observational" is reserved for reports of macro-occurrences that can be so perceived, such as the sounds emitted by some laboratory instrument. On the other hand, statements concerning the distribution of electrons on an insulated conductor or concerning the surface temperature of the sun are sometimes designated as observational, despite the fact that the matters reported are not perceived in the sense indicated. Moreover, the word "theoretical" is frequently used more or less interchangeably with "conjectural," as when a tentative hypothesis as to how a robbery was committed is said to be a theoretical account of the event; and in this case the possibility is not excluded that the conjectured occurrences are also observable in a relatively strict sense of the word, so that they might have been recorded in a suitable set of observation statements. Again, some writers understand by "theory" any essentially general statement, such as "All blue-eyed human parents have blue-eyed children," even if the predicates contained in it designate matters that are usually counted as observable ones. Other writers reserve the word for a system of general assumptions capable of explaining (and perhaps also predicting) the occurrence of a wide range of diverse phenomena, such as the assumptions that constitute Newtonian gravitational theory; and there are some thinkers who use the word rather loosely for the totality of general beliefs, attitudes, or categorial distinctions that make up a Weltanschauung (or, in current idiom, a "conceptual framework" and even a "form of life"). The word "observable" has a comparable diversity of usages, as the above comments suggest.

In any case, it is difficult to formulate with precision the various ways in which the terms "theory" and "observation" are employed. But the senses of these terms that are relevant to the logic of science, and in particular to the central question discussed in this paper, can be illustrated readily enough by typical examples of inquiry in the theoretical sciences. This is not the occasion for an extensive sampling of such specimens, and a single example of what I believe is a representative report of a scientific investigation must suffice. My example is Newton's account of some of the optical experiments he performed in 1666 and of the theory he proposed to explain what he claimed to have observed.

Newton's first letter to the Royal Society containing his "New Theory about Light and Colours" begins with a description of one of his important experiments:

> . . . I procured me a Triangular glass-Prisme, to try therewith the celebrated Phenomena of Colours. And in order thereto having darkened my chamber, and made a small hole in my window-shuts, to let in a convenient quantity of the Suns light, I placed my Prisme at its entrance, that it might be thereby refracted to the opposite wall. It was at first a very pleasing divertisement, to view the vivid and intense colours produced thereby; but after a while applying myself to consider them more circumspectly, I became surprised to see them in an *oblong* form; which, according to the received laws of Refraction, I expected should have been *circular*.
>
> They were terminated at the sides with straight lines, but at the ends, the decay of light was so gradual, that it was difficult to determine justly, what was their figure; yet they seemed *semicircular*.[2]

Newton then went on to report a number of measurements and further observations he made to assure himself that various antecedently plausible sources of the oblong shape of the colored spectrum were not in fact operative. In consequence, he concluded:

> Light is not similar, or homogeneal, but consists of *difform* Rays, some of which are more refrangible than others: So that of those, which are alike incident on the same medium, some shall be more refracted than others, and that not by any virtue of the glass, or other external cause, but from a predisposition, which every particular Ray hath to suffer a particular degree of Refraction.

And he further maintained that

> As the Rays of light differ in degrees of Refrangibility, so they also differ in their disposition to exhibit this or that particular colour. Colours are not *Qualifications of Light*, derived from Refractions, or Reflections of natural

> Bodies (as 'tis generally believed), but *Original* and *connate properties,* which in diverse Rays are diverse. . . . To the same degree of Refrangibility ever belongs the same colour, and to the same colour ever belongs the same degree of refrangibility. . . . The species of colour, and degree of Refrangibility proper to any particular sort of Rays, is not mutable by Refraction, nor by Reflection from natural bodies, nor by any other cause, that I could yet observe. . . .[3]

These quotations will suffice for my purpose. They show beyond doubt that, in describing the observations he made in performing his optical experiments, Newton employed many terms that are "theory laden" in one clear sense of this currently fashionable phrase. Thus, he characterized certain objects as made of glass having the shape of triangular prisms, others as window shutters provided with a small hole, and so on. These predications were not simply reports of sense data "immediately given" to Newton at the time he was carrying on his experiments. On the contrary, the characterizations he used connoted various features of things other than those he explicitly noted as being present—e.g., reflective properties of glass prisms, geometric traits of triangular surfaces, or the opacity of the material out of which the window shutters were constructed. Accordingly, the observation terms used by Newton were not just labels whose contents were exhausted by the directly manifest things to which he applied them, but acquired at least part of their significance from various laws in which they were embedded and which he took for granted. The quotations thus amply confirm what has been familiar to students of human psychology for a long time: that significant observation involves more than noting what is immediately present to the organs of sense; and that a scientific observer comes to his task of reporting his experiments with classificatory schemes representing structures of relations embodied in the flow of events, where the schemes of classification had been acquired in the course of repeated interaction with the environment in ways difficult to recover and enumerate.

However, Newton explained the "phenomena of colours" by a theory that maintained, among other things, that light is composed of distinct "rays," with each ray corresponding to a different color in the spectrum of the sun and having its own inherent degree of refrangibility. But as the above quotations make evident, these theoretical notions so distinctive of Newton's explanation were *not* assumed by him in describing the observations he made when performing his experiments. Moreover, neither his entire letter to the Royal Society nor his much later *Opticks* supplies any ground for holding either that Newton's theoretical explanation of optical phenomena determined (or otherwise entered into) the meanings of the terms he employed in reporting his experimental observations, or that if he had adopted a different explanatory theory those observation terms would have acquired altered meanings.[4] In short, though the predicates which formulated Newton's observations were "theory laden" in the sense indicated, their burden was not the theory he advanced to explain the "phenomena of colours."

A single example doubtless proves little, and I am not supposing that the passages I have quoted from Newton settle the issue under discussion. But in areas of philosophical analysis in which relatively concrete data are not generally used to test some disputed thesis, even a single bit of such concrete evidence may be useful. In any case, the cited example—and only space forbids mentioning others—does confirm the position I am defending, although it certainly does not prove it. The example shows that an experiment intended to ascertain on what factors the occurrence of a certain phenomenon depends can be described so that the statement of the observations made is neutral as between alternative theories which may be proposed to explain the phenomenon, even though the descriptive statement will indeed presuppose various theories, laws,and other background information that are not in dispute in the given inquiry.

2

Numerous objections have been urged against this *prima facie* plausible thesis. In the main, they are directed against the familiar methodological principle that a theory in empirical science must be tested by confronting it with the findings of observation as codified in so-called "observation statements"; and they are at bottom simply variations on the central contention that the theory-observation distinction is untenable, because even ostensibly "pure" observation statements are in fact impregnated with theoretical notions. However, this fundamental criticism is expressed in recent publications in a number of specialized forms,[5] each emphasizing a different facet of the question, and I will therefore discuss some of the more important versions.

One variant of the criticism maintains that the familiar distinction between theoretical and observation predicates (or between theoretical and observation statements) rests on two tacit and closely related assumptions, neither of which is sound. The first presupposition is that theoretical predicates are inherently opaque and hence problematic, so that if their meanings are to be adequately understood they must be explicated in terms of observation predicates, whose meanings are held to be fully intelligible in their own right. But this assumption is declared to be mistaken. For the meanings of observation predicates, so runs the criticism, far from being luminous and unproblematic, are determined by the numerous statements—and in particular, by the frequently comprehensive theories—into which those predicates enter. The point of this objection is stated by Feyerabend in what is perhaps its most extreme form when he asserts that "theories are meaningful independent of observations. . . . It is the observation sentence that is in need of interpretation, and not the theory."[6]

The second alleged presupposition underlying the theory-observation distinction is that theoretical science—and especially theoretical natural science—involves the use of two radically disparate *languages*. One of these is supposedly a self-contained and autonomous language of observation, whose statements deal exclusively with directly observable matters; the other is a language of theory, whose sentences are explicitly about unobservable things and processes, but are nevertheless significant for scientific inquiry only because of their dependent connections with the observation language. But this assumption is also judged to be unsound. For it is not possible to identify in theoretical inquiry two disparate languages of the sort described, so it is argued, but at best only a *single* language between whose various nonlogical expressions distinctions can be drawn, based on differences in their uses or functions.

There doubtless are philosophers who formulate the distinction between theoretical and observation statements in terms of differences between two supposedly distinct *languages*, so that if taken strictly at their word they are legitimate targets for this criticism. However, not all who subscribe to the distinction state it in those terms; and in any case, the double-edged criticism I have been summarizing seems to me to rest on a misconception of the intent of those making the distinction, even when they employ the locution of "two languages" in formulating it. Thus, there is little if any evidence to show that those accepting the distinction generally maintain that there is an "inherent" difference between two sorts of predicates or other linguistic expressions, *irrespective* of the uses to which the expressions are put in various situations. On the contrary, it would appear that the rationale controlling the classification of words like "table" and "electron" as observation terms or as theoretical terms respectively, is that these predicates have recognizably different roles in conducting inquiries (and perhaps even in codifying their conclusions). For example, observation terms are commonly used in experimental investigations for purposes such as the following: to mark off in perceptual experience some spatio-temporally located object or process, that may then be subjected to further physical or intellectual analysis; to characterize an item so identified as one of a certain kind; to describe instruments employed in experiments and what is done with them; to state the outcomes of overt measurements and other perceptual findings, so as to provide instantial premises in inferences involving the application of assumed laws to concrete subject-matters; or to codify experimentally ascertained data, with a view to providing tests for general hypotheses and other statements reached by inference in inquiry.

On the other hand, theoretical terms and statements play quite different roles in scientific investigations, roles that are often characteristically instrumental ones in mathematical physics. For example, theoretical terms sometimes codify highly idealized (or "limiting") notions, such as the notion of an instantaneous velocity or a point-mass, introduced to simplify intellectual constructions or to make possible the application of powerful tools of calculation to the mathematically "imperfect" materials of the natural world. But without going further into such technical matters, it can be said that theoretical expressions have two major functions in scientific investigations: to *prescribe* how the things identified in gross experience with the help of observation terms are to be analyzed (or otherwise manipulated) if the investigations are to be successful; and to serve as *links* in the inferential chains that connect the instantial experimental data with the conclusions (generalized as well as instantial) of inquiry.

This brief account of the functional differences between observation and theoretical expressions is admittedly no more than a sketch. But even this sketch suffices to show that the distinction neither presupposes nor is necessarily committed to the view that observation terms (such as "table" or even "red") are invariably clear, while theoretical ones (such as "electron" or "mass") are inherently problematic. Thus, there are numberless contexts in which neither the meaning nor the applicability of the predicate "table" to some given object raises any issue, as when a waiter in a restaurant is instructed to wait upon certain tables. But there also are situations in which the applicability of the term is uncertain—for example, when the object in question falls into the area associated with the penumbra of vagueness of the word. Analogously, the term "electron" is unproblematic in many contexts (such as the one in which Millikan performed his oil-drop experiment), when the supposed properties of electrons believed to be relevant to the questions under investigation have been articulated with sufficient precision by some theory of electrons. However, both the meaning and the extension of the term may be quite problematic when the accepted theory of electrons does not determine unequivocally whether electrons possess some stated property—for example, when the question is raised within the framework of current quantum mechanics whether, and if so in what sense, electrons are "particles" rather than "waves." In short, the history of science amply testifies that neither observation expressions nor theoretical ones have invariably unproblematic meanings and denotations, and that expressions of both kinds frequently generate difficult questions of interpretation.

One further remark on the "two languages" locution some writers use in stating the observation-theory distinction. It is pertinent to note that the word "language" in this context does not have the sense ordinarily associated with it when, for example, English or French are called languages. For in the present context the word signifies some highly formalized system of notation that is manipulated in accordance with strict rules. Such systems are undoubtedly valuable for achieving the purposes for which they were devised—for example, to make evident certain structural aspects of everyday discourse, to codify standardized procedures for checking the validity of arguments, or to present in precise form certain distinctions. But such systems are in general not suitable (or even adequate) as total substitutes for natural languages in conducting scientific inquiry as well as the daily affairs of life—for example, for communicating the frequently vague ideas that control investigations, or for describing effectively the often complex but frequently unanalyzed operations involved in research. Accordingly, the "two languages" locution should be construed as a pedagogic device for *distinguishing* analytically between important functions of certain groups of expressions employed in the unformalized discourse of inquiry, rather than as a *descriptive account* of two radically distinct languages between which scientists supposedly shuttle. But if this is the proper construal of the locution, the criticism that the observation-theory distinction assumes that observation terms are uniformly unproblematic is as much a gross exaggeration as is the counter claim that there is never a need to analyze theoretical expressions in order to determine their relations to observation terms.

3

Some writers who subscribe to the distinction have suggested that observation terms can be demarcated from theoretical ones with the help of such alleged facts as that the former but not the latter can be predicated of things on the strength of direct observation, or that the former but not the latter have experimentally identifiable instances. However, critics of the distinction deny this to be the case. They maintain that the meaning and use of a new predicate (whether observational or theoretical) cannot be effectively learned or understood on the basis of "direct experimental association alone," and that these things can be accomplished only by presenting the predicate in various sentences containing other descriptive terms whose meanings had been learned previously. But while this claim may be sound, is it really incompatible with keeping the distinction? As was indicated earlier, what is designated by the observation predicates generally employed in experimental research is not simply the ineffable content of any momentary experience, for they signify characteristics that involve more than what may be immediately "given" at some stated time. However, as we also noted earlier, this circumstance is fully consistent with the observation-theory distinction. Accordingly, the criticism under discussion is fatal to the distinction only if observation predicates are equated with so-called "phenomenological" ones—with descriptive predicates that supposedly refer exclusively to what is directly present to the individual who is applying them to matters in his experience. Indeed, if it did presuppose the existence of "purely phenomenological language"—that is, one whose descriptive terms are all phenomenological—the distinction would be quite otiose. For no such "language" is in fact available, and it is a much mooted question whether one is even possible.

Another aspect of the above criticism has also been repeatedly emphasized. This further contention is that the meanings and uses of observation predicates (even of such "basic" ones as "red" and "hard") depend on numerous laws into which the predicates enter, since the laws state, among other things, how the items denoted by the predicates are related to one another. However, the content of these laws is not absolutely invariant, but alters with changes that may occur anywhere in the inclusive network of laws and theories constituting the corpus of scientific knowledge at a given time. In consequence, the meanings of observation predicates are allegedly also modified with changes in that network, so that every attempt to distinguish between observation and theoretical terms is bound to fail. Moreover, critics of the distinction reject as unsatisfactory a frequent and plausible rejoinder to this argument. The rejoinder admits that basic terms like "red" may have "peripheral" uses that may be changed because of changes in the network of laws—for example, the apparent redness of a remote star may come to be regarded not as the color of the star but as an effect of its motion. Nevertheless, so the rejoinder continues, it is difficult to make sense of the supposition that every change in the network invariably affects the "core" meanings of all such terms—for example, in the context of predications concerning things like apples or traffic signals, "red" has an apparently stable meaning that is unaffected by changes in most parts of the network of laws. In rebuttal to these contentions, however, one judicious critic of the distinction has argued that while the stability in limited areas of discourse of the core meanings (or functions) of such predicates as "red" must be recognized, the stability "depends not upon fixed stipulations regarding the use of 'red' in empirical situations, but rather upon empirical facts about the way the world is." The conclusion is therefore drawn that physically possible circumstances can be conceived in which even the core meaning of the predicate would no longer be applicable.[7]

This is a puzzling argument. It is doubtless true, as the argument claims, that "the comparative stability of function of the so-called observation predicates is logically speaking an accident of the way the world is." However, the stability may be genuine and important, without being cosmically necessary or unalterable. Accordingly, the contention involved in the observation-theory distinction that observation predicates do have invariant meanings (even if only in restricted domains of application), is surely not refuted by pointing out that things *could* be different from what they are, and that the circumstances on which the invariances depend *could* disappear. Moreover, advances in science undoubtedly affect the way men talk about the world. But it is not inconsistent to admit that scientific knowledge continues to change, and also to maintain that there is a numerous class of predicates whose meanings or functions in various regions of experience (though demarcated only imprecisely) undergo no significant modifications. It is in part because observation predicates do have such relatively constant meanings that the observation-theory distinction is both warranted and useful.

4

A somewhat different reason is sometimes given for challenging the soundness of the distinction. It is pointed out that predicates commonly classified as theoretical, but which belong to the essential vocabulary of a well-entrenched theory, are often used to characterize matters directly apprehended in experimental situations. For example, certain visible tracks in a cloud chamber may be described as the production of a positron-electron pair; a certain land formation may be characterized as a glaciation; and a person who is observed to walk in a specified manner may be described as having a heart ailment. However, if admittedly theoretical predicates can serve as observation ones, the former cannot be distinguished from the latter on the ground that observation terms characterize what is directly observable while theoretical ones do not. Accordingly, there is no absolute difference in this respect between two classes of predicates, but at best only a gradated one.

It is beyond dispute that theoretical terms are sometimes used in the manner stated. The question nevertheless remains whether this fact impairs the validity of the observation-theory distinction. To begin with, it should be noted that many terms often held to be theoretical are apparently never employed to describe ostensibly observable things. Thus, some occurrences are commonly characterized in terms of their causes, but an analogous account of other occurrences would generally be deemed inappropriate. For example, certain sounds may be described as cannon fire, even though the firing of cannons is not being observed; and the click made by a Geiger counter may be described as the passage of an electron, despite the fact that the passing electron is not seen or otherwise perceived. However, in reporting other occurrences nothing analogous is done. For example, a witness in a criminal court would not be permitted to testify that he *observed* someone fire a gun at the deceased, if all he had actually seen was the victim lying on the floor with blood oozing from a hole in his chest. Nor has anyone seriously proposed, as a description of what is observed in the electrolysis of water, an account of how the electrons in the hydrogen and oxygen atoms are rearranged when the current passes through the electrolyte.

Just why some theoretical expressions are employed to describe observable occurrences (while others apparently never are) is in general unclear as well as controversial. But for many instances of such use of theoretical predicates (as in the case of the term "pair production"), there is a reasonably satisfactory explanation. According to it, the theoretical term in these instances functions as a shorthand formula (based on widely accepted conventions in some branch of inquiry) for characterizing in an effective and discriminating way certain observable but complex features of an experimentally specified occurrence and for distinguishing them from others. For example, the vapor condensations observable in a cloud chamber which, according to accepted physical theory, are effects of pair productions, differ in a number of ways—in shape and thickness, among others—from vapor tracks formed by the passage of alpha particles. However, a long and involved account would be needed if ordinary nontechnical locutions were used to characterize adequately the distinctive traits of the observed vapor condensations in these cases. But since the condensations are associated by current physical theory with the occurrence of certain assumed microprocesses, those acquainted with the theory frequently find it more convenient to use the theoretical predicates rather than the more customary terms of perceptual experience in stating what is observed in cloud chambers.

Accordingly, on this explanation, it is only in a Pickwickian sense that theoretical predicates can be counted as observation terms. Even critics of the observation-theory distinction recognize that descriptions of experimental findings couched in theoretical terms (e.g., that in a cloud chamber pair productions occurred) must be replaced by reports employing predicates of normal perceptual experience (e.g., that white tracks were formed), if the theory justifying that use of the theoretical terms is rejected or even seriously challenged. To be sure, the second description, like the first, asserts more than what is directly apprehended by the experimentor, so that the reason for such a replacement clearly cannot be that no error can possibly occur in using the more familiar observation predicates. The reason for the replacement is that the second description asserts things whose existence is better warranted by the actual evidence than are the matters stated in the initial one. On pain of being totally irrelevant to the purposes of performing experiments, theoretical formulations of experimental findings must cover what can be described in ordinary observation terms; but on pain of being wholly superfluous, the theoretical formulations must also assert, if only by implication, things not stated by the latter description. Therefore, when the theory on which such use of theoretical predicates is based becomes doubtful, and when the theoretical significance of the actual outcome of an experiment is in question, reports of experimental findings stated in noncontroversial observation terms play a crucial role in the conduct of inquiry.

5

The preceding discussion has tried to show that even though observation predicates are in a sense theory impregnated, this thesis is compatible with the observation-theory distinction and is not a good reason for rejecting it. However, some critics of the distinction have objected to it on the further ground that every scientific theory determines the "meanings" of the observation predicates to be used in verifying the theory—that is, the observable data serving as tests of the theory are allegedly interpreted and formulated within a framework of assumptions that form part of the theory being tested.[8] This more radical thesis will now be briefly examined.

On the face of it, if this thesis were correct, arguments for accepting any theory on the basis of empirical data so manufactured would be fatally circular, since nothing would then count as pertinent evidence for the theory which failed to satisfy the relational patterns postulated by the theory. For example, if a moving body could properly be characterized as having a uniform velocity only if no external forces are acting on the body, Newton's first law of motion could never be in conflict with empirical findings, though at the cost of having no factual content whatsoever. However, the history of science provides numerous instances of theories that have been refuted by observational findings. In consequence, the claim that experimental data are always so selected or molded as to fit some assumed theory must be judged as untenable.

This difficulty in the thesis under discussion has been repeatedly noted, and some critics of the observation-theory distinction have tried to meet it. For example, Miss Hesse believes that an important part of the thesis can be saved, if two sorts of terms that could occur in observation statements are distinguished: terms presupposing the *full* truth of the theory being tested, from terms presupposing the truth of *only some* of the laws making up the theory. She acknowledges that if the terms formulating the evidential data for a theory are of the first kind, a genuine test of the theory is impossible; but she holds that if terms of the second sort are used in such formulations, circular reasoning of the type illustrated above does not occur in the test. For example, the truth as well as the meaning of the observation report that a body is moving with uniform speed in a straight line, is said to "depend . . . on the truth of laws relating measuring rods and clocks, and ultimately on the physical truth of the postulates of Euclidean geometry, and possibly of classical optics," all these laws being "part of the theory of Newtonian dynamics." But it is also claimed that since the evidential statement contains only predicates of the second kind, it can serve to test Newtonian theory without circularity.[9]

This contention is patently correct, but the example (and the argument it illustrates) does not confirm anything of moment in the thesis under discussion. It is certainly possible to count the various laws concerning measuring instruments, as well as the laws of Euclidean geometry and even of optics, as parts of Newtonian dynamics. However, these laws do not constitute the distinctive content of Newtonian mechanics, nor is it these laws that are being tested by the observation report mentioned in the example. Thus, to take one instance, Euclidean geometry is known to be compatible with any number of non-Newtonian dynamical systems—the physical laws of Euclid do not entail any of the characteristic Newtonian laws (such as the first law of motion), nor are the meanings of the descriptive predicates occurring in the former determined by Newtonian assumptions. Since it is not Euclidean geometry which is presumably being put to a proof in the example, but rather a law specific to Newtonian mechanics and logically independent of Euclid, it is hardly surprising that no circularity is involved in the test. Moreover, the argument is incompatible with the view, integral to the radical thesis being discussed, that if a theory is changed or completely replaced by another, all observation terms used to state the evidential data for it necessarily change their meanings also. For the argument is built on the premise that some observation predicates used to state the evidence for a theory do not presuppose the truth of *all* the laws making up the theory. Such predicates can therefore continue to serve as observation terms with unchanged connotations for a different theory, obtained from the original one by changing one or more laws of the latter that are not presupposed by the predicates. In short, the argument makes coherent sense only on the assumption—associated with, but not distinctive of, the observation-theory distinction—that though the meanings of observation terms are at least partly determined by the laws into which the terms enter, the laws do not form a single, monolithic system of statements no two of which are logically independent.

This assumption also underlies Miss Hesse's attempt to counter a further objection to the thesis under consideration: if the meanings of the descriptive terms in observation reports were indeed fixed by the theory for which the reports can serve as evidence, the very *same* reports could not confirm each of two fundamentally *different* theories, nor confirm one and disconfirm the other. But if this were so, it would be impossible in principle ever to decide between ostensibly competing theories on the basis of observational findings—a conclusion that runs counter to actual scientific practice. But according to the proposed answer to the objection, this conclusion follows only in certain cases. If two theories really have no concepts in common, the conclusion is indeed inescapable. However, even theories that differ profoundly in their foundational assumptions and their implications may nevertheless contain "some hard core predicates and laws which they both share." For example, despite the far-reaching differences between Newtonian and Einsteinian dynamics, both employ such important predicates as "acceleration of bodies falling near the earth's surface" and "velocity of light transmitted from the sun to the earth," and they also have in common a number of laws into which these predicates enter. In consequence, so the argument continues, if an observation report contains predicates occurring in laws that belong to both theories, the same report can be used to decide between the different theories. But here again the proposed resolution of the difficulty must grant the crucial point, vigorously denied by some critics of the observation-theory distinction, that the meanings of observation predicates are not completely determined by some given theory for which the predicates serve in formulations of evidence. The argument makes sense only on the assumption that though the meaning of a predicate P may be fixed (perhaps only in part) by some law L that belongs to each of two different theories T_1 and T_2, the meaning of P cannot depend on every other law in T_1 or every other law in T_2 if an observation report containing P is to test both theories. For otherwise, what the report would be saying when testing T_1 is different from what it would be saying when testing T_2, so that the *same* report would not (and indeed could not) be used to help decide between the two theories.

Science has often been compared with myth. Scientific theories, like many myths, are attempts to account for what happens in various sectors of nature; and, also like myths, they are works of the imagination that carry the impress of an enduring human condition as well as of variable special circumstances. It is therefore not surprising that scientific theories and myths have many common traits. Indeed, as Heinrich Hertz suggested in so many words in his treatise on the principles of mechanics, every system of symbols that makes up a scientific theory is bound to have components which can appropriately be called "mythical." For even when a theory is adequate to the facts for whose explanation it was devised, it will inevitably possess features that represent nothing in the subject matter, but reveal instead something about the capacities and preconceptions of its creator.

However, a controlling assumption underlying the development of science since ancient times is that properly conducted inquiry can arrive at explanatory theories that are not wholly mythical. If this assumption is sound, science and myth must not only be compared, but also contrasted. Such contrast has been frequently made: for example, by Einstein in the passage quoted at the outset, or by George Santayana when he declared that science differs from myth "insofar as science is capable of verification." To be sure, the probative value of verification has been seriously disputed throughout the centuries. And if the doubts about it raised by some recent commentators on the logic of science are well-founded, the belief that inspired the creators of modern science—that the truth about things can be found out by inquiry—is itself a myth. The critique in this book of some currently influential arguments eventuating in a sceptical relativism is intended to show that those doubts are not warranted, that this ancient belief is still tenable, and that scientific theories cannot be equated on principle with old wives' tales.

3/ Some Reflections on the Use of Language in the Natural Sciences

One of the ideals of the natural sciences, splendidly if but partially realized in some of them, is the discovery and formulation of a structure common to diverse processes, in terms of which the behaviors of qualitatively heterogeneous things can be systematically understood and controlled. To take just one series of examples, consider such superficially unrelated phenomena as the behavior of a tear rolling down a human cheek, the rise of sap in the stems of plants, the motion of a planet through the stations of the fixed stars, the action of a violin or a trumpet when producing sounds, or the boiling of a kettle of water standing over a flame. In spite of the manifest and irreducible differences between these activities, modern science has taught us to recognize in them a common pattern. These qualitatively incommensurable processes are shown in the science of mechanics to possess common features and an identical mode of action; and, accordingly, alterations in them which may at first sight appear to be randomly spontaneous are disclosed to depend upon a repeatable structure of relations.

Such pervasive structures are formulated by the principles of the natural sciences. And it is a truism that the content of such principles is not anything which can *literally* be seen, heard, or felt—it is not a matter for visual, auditory, or tactual experience. For the common pattern of physical changes can be exhibited only symbolically, through the medium of language. It is language and language alone which renders what is possessed in common by existentially distinct processes, and which, as the familiar phrase puts it, reduces heterogeneous things to a common denominator. Just as in the economic market money is the medium in terms of which qualitatively incommensurable goods can be systematically compared, so more generally language is the means by which existentially unique processes are equated, ordered, and understood.

The rôle of language in science has been often recognized; indeed, according to a famous and influential dictum, science is nothing but a well-made language. This dictum has long inspired the conviction that our ignorance of nature's ways can be dissipated through the institution of an appropriate system of notation; and even in our day, a panacea for almost every variety of social evils has been found in certain proposed alterations of current linguistic habits. However, the dictum rests upon a conception of human psychology which is no longer credible; and, in any case, science certainly involves more than the institution of an adequate method of symbolic representation and manipulation. Nevertheless, though Condillac's dictum states something less than the whole truth about science, it does express a fraction of it. Many of the distinctive traits of the scientific enterprise are embodied in its mode of using language, so that the character of its symbolism is one mark which differentiates science from everyday inquiry and common-sense beliefs. The following somewhat desultory reflections are intended to call attention to some features of the natural sciences—in particular, to the deliberate diminution of *vagueness* and of *unspecific usage* in their symbolic structure—features which contribute to the achievement of an adequate symbolic rendering of pervasive structural patterns in physical processes.

1

An arresting and frequently disturbing feature of everyday language is the vagueness of its words and phrases. This phenomenon was well known to the ancients, and they constructed instructive puzzles based upon it. The following is a modernized version of one of their paradoxes. We commonly distinguish men that are *bald* from those that are not. But under what circumstances is a man to be so designated? It seems at first sight plausible to say that a man is bald under the following two conditions: (*a*) if a man has *no* hairs on his head; and (*b*) if he has $n + 1$ hairs, *provided* that he would be bald were he to have just n hairs. For surely, so it might be argued, the addition of a *single* hair does not transform a bald man into one who is not bald. However, a moment's reflection shows that were we to accept *both* parts of this attempted definition of "baldness," *all* men with any finite number of hairs on their heads would have to be called bald—a patently absurd outcome. The paradox is immediately resolved, however, if we note that the word "bald" as ordinarily used is decidedly vague, that the men who are called "bald" are not precisely distinguished from those not so described, and that therefore the second part of the proposed definition must be rejected if the customary usage of the word is to be retained.

This illustration is trivial and perhaps amusing. However, it is cited not for entertainment, but for the sake of several important observations which may be attached to it.

(1) It is essential to note that the vagueness of language does not arise from lack of *relevant knowledge* on the part of those who employ it. Thus, even if it were generally known exactly how many hairs each man possesses, the word "bald" in its current usage would still be vague. Vagueness has its source in the fact that *habits of usage* are not *determinate* in applying linguistic expression to certain ranges of objects. Stated more explicitly, a word is vague if the things *habitually* characterized by it are not sharply distinguished from those things which are *habitually* denied that characterization—with the consequence that for some things falling within its range of application the use of the word is indeterminate. The occurrence of vague expressions thus testifies at once to the existence of experienced *continua* of qualities and dispositions, and also to the absence of *fixed habits of discrimination* between segments of such continua. Accordingly, the vagueness of linguistic terms can be diminished (though perhaps never completely eliminated) only through the deliberate *reconstruction* of language habits so as to institute sharper conventional delimitations of qualitative differences.

It is not always desirable or useful to diminish the vagueness of language. In poetical discourse, vagueness is often an advantage, rather than a defect. And in the context of ordinary affairs, nothing worthwhile would be gained by making such words and phrases as "bald," "moving quickly," or "sticky" more precise than they in fact are. On the other hand, it is obvious that vague language can be a serious hindrance to the execution of social policy as well as to theoretical research. The vagueness of the adjective "adult" in its ordinary usage is intolerable for the purposes of law, and it is redefined in the latter context in terms of chronological age. Again, though in many affairs of life we manage very well with such vague phrases as "moving quickly" or "sticky," we would be utterly hamstrung in physics and chemistry if we possessed no more precise ways of formulating the behavior of things than by saying, for example, that light moves very quickly or that some liquids are more sticky than others. In these sciences we replace vague expressions by highly precise ones, such as "velocity" and "degree of viscosity."

It is worth at least passing mention that precision in the usage of language is deliberately instituted in the sciences in order to make possible the formulation of *general laws,* and thus to render symbolically the *structures* of mutual dependencies of things and their properties. The diminution of vagueness is not undertaken simply for the sake of the formal ideal of an absolutely precise vocabulary. It is idle pedantry to demand the same degree of precision in *every context,* independent of the special objectives for which a symbolic system is required. On the other hand, it is often a sign that inquiry is achieving a more subtle and comprehensive organization of knowledge if vague terms must be replaced by more precise ones. For an increase in the precision with which language is used indicates that differences hitherto unnoticed or regarded as unimportant are being recognized as determining factors in the structure of physical processes. Precision of language is achieved through an explicit formulation of rules of usage; and there are many special techniques which may be employed to attain this end. In the physical sciences, counting and measurement are the preeminent, but by no means exclusive, ways for instituting less vague language. It is indeed a naive conception which identifies precision with quantitative determination. That supposition has led many workers in various fields of social studies to become furiously and sometimes exclusively occupied with measuring anything for which they can devise a scale—as if quantitative discrimination were the primary objective of knowledge, and as if a highly precise vocabulary were sufficient to constitute a science.

(2) It follows from this account of vagueness that whenever a term in current usage is redefined so as to be less vague, a *new* usage is introduced for that term, and a familiar concept is replaced by a *fresh* one. This simple point helps to explain why common-sense notions

are so difficult to analyze, why there is the widespread conviction that intellectual analysis mutilates what it analyzes, and why, in particular, the formulations of the physical sciences are so often believed to be unfaithful to the vivid, concrete reality of familiar things and processes.

Suppose, for example, that one wished to analyze the meanings of such expressions in daily usage as "justice," "personality," "disease," "heat," "necessity," or "cause." These words are unquestionably vague, to say nothing of their being ambiguous. But an analysis of the meaning of any one of them, "justice" for example, is usually required to formulate the common structure of the situations to which the word is applied, and thereby to express the meaning of the word with the help of a longer but synonymous expression. However, if the word is vague, not only is the range of its application indeterminate; it also is most unlikely that any expression or combination of expressions in the language will convey the customary sense of the word—for the penumbra of indeterminateness associated with one set of expressions in the language does not in general coincide with the penumbra associated with another set. Accordingly, when a definition for a term in common usage *is* proposed, the proposal will usually involve a *revision* or *reconstruction* of the initial usage of the term in question, if only because the defining phrases will tend to restrict its original vagueness. It is therefore perhaps inevitable that most proposed analyses of the meanings of linguistic expressions in common usage tend to produce a feeling of discomfort in their users, and to give rise to the general sense that definitions "leave out" something ineffable but important, or that analysis corrupts that which it pretends to analyze. It is also understandable why certain notions, like *goodness* and *causality*, are so often claimed by philosophers to be *inherently* unanalyzable: for if the adjective "good" is vague, no other word or phrase (in terms of which an analysis might be stated) will be strictly equivalent to it, simply because of the unique area of vagueness associated with each expression in the language.

The feeling that something vital is lost in the process of analysis and definition is accentuated when words like "heat," "combustion," or "disease" are taken over from everyday language and assigned a precise, technical usage in some special department of science. For as a consequence of such technical refinements, the customary human associations and overtones of such terms are eliminated; and in its new usage, the range of application of a word as commonly employed coincides only in part with the new range specified for it. Those for whom everyday language provides the only adequate measure of reality rather naturally conclude, therefore, that the general statements of the natural sciences—since usually they cannot be translated without remainder or irrelevant additions into the language of daily life—distort the true character of their subject matter. According to Bergson, for example, there is something "artificial in the mathematical form of a physical law and consequently in our scientific knowledge of things"; and he assigns to philosophy the task of transcending "the geometrizing pure intellect" by reintroducing us "into that more vast something out of which our understanding is cut off and from which it has detached itself" through its symbolic rendering of natural processes. One motivation of Bergson's philosophy thus seems to be rooted in his felt disparity between the vague discourse of daily life and the relatively more precise language of science.

But unless one adopts the absurd conception of knowledge, according to which the goal of thought is an intuitive, empathic identification with the things to be known, there is very little substance in Bergson's critique of scientific knowledge as "abstract," and of language and symbolism as impediments to knowledge. It is undoubtedly the case that thought and analysis are selective activities; and perhaps nothing shows so clearly their selective character as the constant reconstruction which language must undergo in the interest of precise formulation. But since reflective analysis does not aim at a reduplicative formulation of whatever exists, such selection and reconstruction do not entail any scepticism concerning the ability of scientific inquiry to achieve its proper function. In particular, although when the vagueness and functions of language are borne in mind it becomes incredible to suppose that the *structure of any language* reflects (or is isomorphic with) the *structure of what it expresses,* the symbolic systems of the sciences are capable of rendering the mutual dependencies of things in more inclusive and subtle ways than the language of daily use can manage to do.

(3) One final point needs to be made in connection with vagueness. Systematic formulations are normally achieved in the natural sciences, not by employing terms *referring* directly to matters of *familiar experience,* but by introducing highly precise expressions de-

fined within the framework of theoretical assumptions; and such expressions often signify properties of things that are specified as the *limiting* termini of nonterminating serial relations. Accordingly, expressions are instituted which, simply as a consequence of their extreme precision, appear to represent nothing identifiable in the subject-matters explored. It follows that statements containing such expressions are *prima facie inapplicable* to anything on land or sea.

An example will help to make more evident the problem that is thus engendered. In ordinary affairs we distinguish between bodies at rest and those in motion; and we also distinguish between different speeds with which bodies move. But the term "speed" is too crude a characterization for the purposes of physical science, and as everyone knows physicists distinguish between "velocity" and "acceleration" in order to formulate the laws of bodies in motion. Now consider the expression "moving with constant velocity"—to what class of motions is it to be applied? The velocity of a body is sometimes defined as the time-rate of change of position of a body along a straight line; and a body can be characterized as moving with constant velocity only if the ratio of the distance a body moves to the time of its motion is constant during some interval of time. However, in saying that the velocity of a body is constant, we are asserting that the distance covered during *each unit of time* remains the same—when the unit of time, which for example may be an hour, is arbitrarily specified. It is clear, therefore, that "constant velocity" so construed represents an *average value* of a ratio—since, though the distance covered by a moving body may be constant when one specified unit of time (say, an hour) is taken, the ratio of distances to times may vary if a *smaller* unit of time (say, a minute) is selected. But in order to be able to state the laws of motion in a perfectly general manner—so that they may be applicable to all types of physical phenomena and however the unit of time is chosen—the notion of the velocity of a body "at an instant" is introduced. The expression "velocity at an instant" thus represents a culminating redefinition of an initially vague term used to describe the motions of bodies. On the other hand—and this is the point around which serious methodological problems cluster—no overtly identifiable motions of bodies can be characterized as having a "constant instantial velocity." The net outcome thus seems to be that in redefining familiar distinctions so as to make them highly precise, certain "limiting concepts" are introduced into the language of science which *apparently* have no pertinent use in connection with empirical subject matter. In consequence, many thinkers have been led to suppose that such statements are in some sense "self-authenticating," that they require no experimental warrant, and that they are guaranteed entirely by the symbolic operations to which they are subject.

But this supposition falls rather generously short of the truth, and the difficulty which has been briefly indicated is not a fatal one. For when the laws of the natural sciences are formulated with the help of "limiting concepts," those theoretical formulations are always required to be supplemented by *various specific rules of application*—rules which coordinate the highly precise terms of theory with *relatively vaguer terms* of overt practice.

The use of language in the natural sciences thus exhibits a characteristic rhythm. Linguistic expressions which are initially vague are replaced by others less vague; and in order to achieve a maximum of generality with an economy of formulation, various "limiting concepts" having a minimum of vagueness are finally instituted. On the other hand, in order to apply those formulations to overtly experimental material, the ideal of absolute precision must be relaxed: the relevance of theoretical formulations to matters of direct experience must be indicated through the use of language *less precise* than the language of theory, though usually *less vague* than the language of everyday discourse. This ebb and flow in the adoption of various degrees of precision is one of the most arresting features in the scientific use of language; and a central task for any philosophic critique of the "abstractions" of science is to exhibit the functions of each stage in this series of transformations. For such a critique must aim to prevent the hypostatization or reification of instrumental functions into mysterious, inaccessible agents, and to show how scientific language, in conjunction with overt experimental procedure, can render, but without illusion, the pervasive interrelations of natural processes.

2

This brings me to a second feature of language which, though closely related to the phenomenon of vagueness, warrants special attention. Consider, first, an example taken from ordinary discourse.

Children learning to use English may be taught that the word "valley" refers to depressions between elevations of land. Presently, however, they find the word employed in such a sentence as "The boat was invisible in the valleys of the sea"; and they discover that its initial range of application is extended to include depressions between elevations of water as well as of land. But subsequently they may come upon such statements as "He dwelt in the valley of the shadow of death"; and though they may at first be perplexed by this curious use of the word "valley," they become sensitive to certain analogies between its use in this new context and its previous uses. It is quite unlikely that most children or their elders ever learn to state clearly the conditions under which the word may correctly be used in this novel way; nevertheless, people develop unformulated habits governing this extended employment of the word, and can distinguish between true and false statements containing the word in this usage.

This example makes evident that an expression having an established usage *in one determinate domain* may acquire radically new though related ones through *a gradual shift* in the context of usage. Such new senses for expressions are ordinarily not established through any explicit formulation of rules of usage; and in many cases, certainly, no serious confusions arise as a consequence. Nevertheless, the uncritical extension of the range of application of an expression, even when guided by analogies, is a common conviction that since the *same expression* is employed throughout a series of contexts, there must be a *generic meaning* common to all its uses—as though, to use a happy image of Wittgenstein, continuous fibers must stretch throughout the entire length of a hawser in order that a ship be securely tied to its pier. It fortifies the tendency toward hypostatization, toward the introduction of *substantial but essentially mysterious agents and entities* for which the *relational-terms* of scientific discourse are taken to be surrogates. It easily produces the obscurantist conceptions of nature which are a consequence of construing the meaning of statements occurring in one context of operations in terms of meanings possessed by statements in quite different contexts. And it leads easily to the facile but uninformed scepticism expressed by a well-known novelist, according to whom modern science "is simply incomprehensible. Every statement is a paradox, and every formula an outrage on common sense"—a conclusion generated entirely by a failure to attend to the meanings of scientific statements in terms of the specific habits or rules of usage with which they are associated.

This partial catalogue of possible consequences of the uncritical extension of language represents dangers encountered in the scientific as well as in the daily use of language. They are dangers which can be overcome only by carefully *specifying* the rules of usage for linguistic expressions in *each context* of their application, and by resolutely interpreting the meanings of statements on the basis of such rules. The ideal end at which science aims in the construction of its language is the exclusion from its symbolism of all expressions which have no *specified* usage. And the substance of the methodological program currently known as "operationalism" is that of supplying such *rules* of usage for expressions not already possessing them. In the remainder of this chapter I wish to cite a few examples from the sciences which illustrates how some of the cited dangers have been incurred, and how, in some cases at least, they have been overcome.

(1) Algebra, as is well known, was first developed as a generalized calculatory device for handling arithmetical subject matter—the integers, their ratios, and their relations. But presently the algebraic algorithm produced certain curious combinations of symbols—obtained in the process of solving equations—which obviously did not "stand" for any of the familiar arithmetical numbers. Accordingly, on the assumption that every combination of symbols obtained by *apparently* employing the accepted rules of symbolic manipulation *must* stand for *something,* mathematicians thought it necessary to postulate various new kinds of "mathematical entities." Like a veritable sorcerer's apprentice, the machinery of algebra thus produced the negative numbers, the imaginary numbers, and the irrational or surd numbers; and the platonic mathematical heaven became thickly populated with these semimysterious entities, conceived as subsisting in their own right antecedently to the symbolic operations which merely called them to men's attention. However, a careful study of the procedures involved in the postulation of these entities—a study which has required something like 200 years of devoted thought to complete—has finally shown how entirely gratuitous has been the postulation of these new kinds of numbers as substantial beings. By specifying fully the rules of usage governing expressions ostensibly representing these hypostatic beings, modern research has made plain that the

hypostatization depends on *covertly and illicitly* extending the use of certain symbolic operations from domains in which they had been well-defined into domains for which they had *not been defined at all.* The problem which thus faced mathematicians was that of constructing *definitions* for certain symbolic operations, and of finding fruitful *interpretations* for symbolic complexes obtained by these operations. And when appropriate definitions and interpretations were introduced, it was seen that the curious combinations of symbols which previously were taken to represent self-subsistent but not quite intelligible platonic entities were symbolic of *identifiable relations between identifiable things*—of operations between integers in some cases, of operations upon geometrical figures in others, and so on. It is worth noting that through this analysis the power of algebraic procedure has been increased rather than diminished, while at the same time the functions and limitations of symbolic manipulations have been clarified.

A similar moral may be drawn from the history of higher mathematical analysis. Every adult understands what is meant by the word "sum" in such contexts as "the sum of 1 and ½" or "the sum of 1 and ½ and ¼," where only a finite series of numbers are to be added. Does the word mean anything when it occurs in such contexts as "the sum of 1 and ½ and ¼ and ⅛ and so on to infinity?" Mathematicians at first supposed that in talking of the sum of an infinite series of numbers they were employing the word "sum" in its *customary* sense and in accordance with *customary* rules of usage already established for finite series. But they presently discovered that in simply extending the use of the word to infinite series fatal contradictions arose; and they finally recognized that the word "sum" for infinite series was *not defined* by the rules which govern its customary use. In other words, it does not follow that because the expression has a determinate meaning in contexts of adding a finite series of numbers, it has a determinate meaning for combining an "endless" sequence of numbers. Mathematicians finally formulated rules for combining such endless sequences; but they still take inordinate pains to note that such sequences possess no "sums" (for this word is reserved for the addition of finite series of numbers), though in certain cases they possess something called "the limit of the sum." Moreover, the expression "limit of the sum," if construed in terms of the rules which define it, can not be taken to represent an unknown "something" transcending the series of numbers; it represents in a compact fashion a complex *pattern of relations* between members of the series of numbers. In this instance, also, a careful specification of rules of usage makes *ad hoc* postulation of "entities" quite gratuitous; and it makes evident the *dependence* of apparently self-subsistent mathematical subject matter upon the process of *symbolic construction*.

(2) Consider next some examples from physics. In traditional formulations of the laws of mechanics, the motions of celestial bodies are usually explained in terms of a "force of attraction" acting between each pair of bodies. However, the word "force" was taken over into theoretical mechanics from the language of daily practice; and it was perhaps inevitable that explanations of motions in terms of forces should therefore be associated with the notion that strains and tensions are involved in the behavior of celestial bodies which are in some way analogous to those directly experienced when human muscles are exercised. But when mechanical forces are conceived in this manner, they become inaccessible to direct or indirect observation: they are then mysterious *agents*, whose operations become matters *requiring* explanation instead of serving as explanations for other things. Accordingly, the view became widespread—among practicing scientists as well as among laymen—that although we may know what attractive forces *do*, we are quite ignorant, and must forever remain ignorant, of what they "really" *are*. But this hypostatization of forces into agents was found to be altogether unnecessary when physicists finally proceeded to formulate carefully the rules in accordance with which the word "force" needs to be employed in mechanics, and in terms of which the meaning of the word needs to be understood. It then became evident that the word is a short-hand symbol for a complex system of *relations* between the masses and relative accelerations of bodies and their distances from one another. This analysis did not show, as is sometimes maintained, that the word "force" is a "mere word," an empty sound; it did reveal the office of the word in rendering systematically and symbolically a pervasive structure of physical behaviors, and it did make clear that this office could be exercised without converting patterns of behavior into unidentifiable agents *producing* that behavior.

An interesting variant of the above example is found in the history of the term "energy" as employed in physics. The beginning student in physics is usually told that by the "energy" of a physical system he is to understand the capacity of the system to do work; and he soon learns that a certain function of the mass and velocity of a moving body (its kinetic energy) is a measure of this capacity. It also turns out that when the equations of motion for an isolated physical system are transformed in certain familiar ways, a certain quantity remains constant—a quantity which is the sum of two others, one of them being the kinetic energy of the system. This constant quantity is called the "total energy" of the system; and the second of the two terms of which it is the sum is then rather suggestively designated as its "potential energy." In physics, as in the case of algebra, the algorithm of symbolic manipulations thus gives rise to new combinations of the elementary symbols; and, as in the case of algebra, these new symbolic complexes are often taken to denote something substantial but inaccessible underlying physical processes. Thus, the statement that the total energy of an isolated mechanical system remains constant throughout any transformation has frequently been construed to mean that a mysterious something called "energy" now takes the form of kinetic energy, now the form of potential energy, although it maintains its substantial identity whatever its accidental phenomenal form may be. It is not difficult to see why physics appears as paradoxical and an outrage to common sense, when this mode of interpreting physical principles is applied to such a statement as that the mass of a physical system may be transformed into energy. However, the air of paradox disappears entirely when words like "mass" and "energy" are understood in terms of the specific rules governing their use. It then turns out that the expression "energy of a system" does not represent some self-identical but hidden "entity" or "agent"—but is an economical device for formulating a constant *pattern* of behavior exhibited by physical bodies.

But perhaps the by now classical illustration of the rôle of explicitly formulated rules of usage in the development of physical theories is provided by the relativity theory. Most people are confident that the statement "Two events are simultaneous" has a determinate and unambiguous sense; and there is indeed no doubt that within the context of everyday practice the word "simultaneous" has such a meaning. However, the customary use of the word in connection with immediately practical matters was gradually extended so as to cover the temporal relations of events occurring under highly complex conditions. And shortly after the turn of the present century it was recognized that for this *extended* application, the word was *not defined* at all, since it has *no specified* usage. The rest is well-known history: rules of usage for this word were carefully instituted, and the physical theory of relativity was developed. The construction of this theory involved much more than the institution of definite rules of usage for expressions hitherto employed without a specified meaning. But the development of that theory does exhibit the realization of one of the ideals of scientific language—namely, the exclusion of a symbolic construction for which no determinate conventions of usage have been specified.

(3) Let me now briefly indicate the failure of this ideal in some social studies. Historians, sociologists, and political theorists often employ the language of physics in formulating the results of their researches. Thus, their writings contain such phrases as "the force of personality," "the intellectual energy of a community," "the friction involved in competitive behavior," "the acceleration of social change," or "the osmotic pressure of cultural infiltration." Such expressions are frequently innocent of pretentiousness; they are literary devices for stating in arresting language what everyone understands. And if they deceive no one, it is misplaced hygienic zeal to demand the elimination of such metaphorical figures of speech from the language of social analysis. However, the use of the language of physics in this context is not always innocent. Such language is sometimes employed to lend an air of profundity and precision to sociological statements—traits not always supported by a closer analysis; and, indeed, highly debatable if not irresponsible opinions are passed off under the mask of a language for which no determinate rules of usage are available. In discussing the rôle of religious thought and of economic institutions in the development of modern capitalistic society, for example, it is utterly futile to maintain that one of these has been a "greater force" than the other in determining the character of our society—until the rules of usage for the expression "greater force" have been made definite. In a recent and widely hailed treatise on sociology, the author paraphrases the Newtonian laws of motion for societal changes, and submits that sociology has finally been placed on a firm "scientific foundation" in consequence. But the writer manifests no sense of obligation to specify how phrases such as "social force" or "acceleration of social change" are to be employed. For ex-

ample, nothing is said as to how such forces are to be compared, what it means (if anything) for such forces to be combined in accordance with the parallelogram law of addition, and the like. In this instance, certainly, the ideal of instituting *specified* usages for all symbolic constructions, an ideal so faithfully pursued in the natural sciences, has not even been envisaged.

I do not for a moment wish to deny the suggestiveness and stimulus which comes from the extension of linguistic expressions from an initial domain of application to new fields. On the contrary, I have tried to indicate how, in the mathematical and physical sciences, such extensions, provoked by vaguely felt analogies and frequently generated by the automatism of a symbolic system, may yield more general and more efficient symbolic instruments. Nevertheless, such extensions in usage must always be critically scrutinized and supplemented by explicit rules of usage—if the price in initial confusion that must often be paid for such advances is not in the end to outweigh the advantages that may accrue from them. Symbolic operations are never self-authenticating in empirical inquiry, and no error is more serious or more insidious than that which arises from the tacit supposition that linguistic manipulations guarantee the representative character of the symbolism employed.

4/The Quest for Uncertainty

The thesis, central to Cartesian rationalism as well as to sensationalistic empiricism, that genuine knowledge of matters of fact is impossible unless at least some statements can be asserted as true with incorrigible finality, has been the target of repeated criticism. The thesis continues to be defended in various forms, despite such criticism; but the claim that some factual propositions are indeed known with "absolute" certainty can no longer be plausibly supported, as it once was, by introducing as evidence the history of modern natural science. When Charles Peirce was developing his "contrite fallibilism" during the latter part of the previous century, it was commonly assumed even by outstanding physicists such as Kelvin that the "universe [was already] entirely explained in all its leading features; and that it [was] only here and there that the fabric of scientific knowledge betray[ed] any rents."[1] However, the climate of opinion illustrated by this assumption has been sufficiently modified, at least partly as the result of basic changes in theoretical physics and biology since the beginning of the present century, so that the accounts which the natural sciences give of the various processes investigated by them no longer seem convincing examples of indubitably final knowledge.

In any event, Peirce's doctrine of fallibilism (like Dewey's subsequent critique of the quest for certainty which developed some of Peirce's ideas) sought to undermine philosophic conceptions of knowledge that appeared to him to "block the way of inquiry" by generating such claims to finality. His analysis of the logic of science attempted, among other things, to remove what he described as four "venomous errors" obstructing the path of science and thereby the growth of knowledge: the error of assuming that the self-corrective method of science (regarded by both Peirce and Dewey as the most assured way yet devised for ascertaining "by reasoning how things really and truly are"[2]) can ever yield "absolutely certain" conclusions which are by their very nature exempt from correction in the light of further inquiry; the error that some things "can never be known"; the error that some facts of science are "basic, ultimate, independent of aught else, and utterly inexplicable"; and finally, that error that certain "laws" or "truths" currently accepted in the sciences have received their final and perfect formulation.[3]

It is difficult to assess the extent to which the criticisms of either Peirce or Dewey have been effective in altering the climate of opinion in which these assumptions were not commonly questioned. But however this may be, it is widely though not universally acknowledged today that *all* cognitive claims in science—no matter how securely grounded they may appear to be, and whether they are about particular occurrences or about relations of dependence between inclusive kinds of things—are *in principle* subject to further scrutiny and possible revision, so that in this sense the fallible character of scientific inquiry has become a commonplace. However, this commonplace is being currently transformed into a paradox. For there is a growing tendency among scientists as well as philosophers to argue from the fallibility of scientific inquiry to a denial that claims to knowledge are ever warranted, and to describe the enterprise of science as if it were a quest for uncertainty. For example, Max Born, one of the architects of quantum mechanics, made the surprising confession that after a lifetime in physics he was convinced that his early belief in the superiority of science over other modes of thought as a way of achieving objective knowledge of the world was a self-deception. Born's present evaluation of science represents an extreme reaction to certain features of quantum theory; and although his views are shared by other distinguished physicists, I will not stop to consider them here. My aim in this chapter is to examine several claims, advanced by a number of philosophers on grounds different from Born's, that raise doubts—by implication when not explicitly—whether, because of its fallibility, scientific inquiry ever yields assured knowledge of the world.

1

A vigorous exponent in recent years of one version of fallibilism has been Karl Popper, and he has also championed the view that the scientific enterprise is a persistent search for the truth about the constitution of things. In particular, in opposition to certain forms of positivism and pragmatism, he has been steadfast in maintaining that scientific theories (which he equates with systems of strictly universal statements) are not merely instruments of calculation or classification, but have an informational content, and provide *explanations* for various features of nature. Nevertheless, Popper denies that science achieves genuine *knowledge* of any matters of fact, and claims that on the contrary every assertion of science is "merely a guess" or "conjecture."[4] Indeed, science is said by him to be "not interested" in establishing its conjectures, "anticipations," or theories as either "secure, certain, or probable . . . [but] only in criticizing them and testing them."[5] The method of research in science, he declares, is directed "not to defend [the conjectures or 'anticipations' that have already been put forward, but] on the contrary . . . to overthrow them. . . . We try to prove that our 'anticipations' were false—in order to put forward, in their stead, new unjustified and unjustifiable anticipations."[6]

It is perplexing to find that what is ostensibly a defense of the notion that scientific inquiry is a search for answers to problems which initiate it should in effect maintain not only that science never really finds answers to its problems, but also that the search for them is necessarily unsuccessful. How does Popper arrive at this strange conclusion? Before discussing this question I must first explain some of his terminology.

Fundamental to Popper's analysis of the logic of science is the distinction he thinks it is essential to draw between statements that have an *empirical* content (and which therefore belong at least potentially to some empirical science) and so-called "pseudo-empirical" or "metaphysical" statements that lack such content. He rejects the suggestion that the required demarcation be made in terms of the *verifiability* of statements—on the ground that since universal statements are patently not *completely* verifiable, the suggestion would exclude every theory from the category of scientific statements, and would thereby also exclude the search for explanations as a proper task of science. According to the criterion of demarcation he adopts, a statement is empirical only if it is *falsifiable,* that is, only if it is logically possible to refute it by experience. However, it is essential for an understanding of Popper's argument to follow him in his somewhat more precise formulation of this criterion.

Popper calls a statement "basic" if it has the form of a singular existential statement (i.e., There is a so-and-so in the spatiotemporal region k) which says that some observable thing or event occurs at some designated place during a given period of time. For example, "There is a penguin in the Bronx Zoo this evening," and "Between 3 and 4 P.M. yesterday there was a pendulum oscillating with a period of one second in the Pupin Building," are both basic statements. Assume next that the notion of "class or universe of all possible basic statements" can be well defined. Then a statement is falsifiable (and hence empirical) if it divides the universe of basic statements unambiguously into two nonempty subclasses—the subset whose members are *incompatible* with the given statement (which are therefore called its "potential falsifiers"); and the subset whose members are *compatible* with the given statement. An empirical statement is said to be *falsified* or *refuted,* as distinct from being merely falsifiable, if one or more of its potential falsifiers is accepted as being in agreement with observation. A statement is also said to be "testable" if it is falsifiable; and to test a statement is to try to falsify it. On the other hand, a statement is said to be "corroborated" or "confirmed" if it is compatible with accepted basic statements, and if in addition a set of basic statements (which are accepted as the result of "sincere" attempts to refute the statement) is logically *derivable* from the statement in conjunction with other accepted basic statements.

It will be evident that a statement which is compatible with *every* basic statement, so that the class of its potential falsifiers is empty, is making no empirical assertion whatsoever. An empirical statement thus *excludes* certain logical possibilities from the realized ones, and its content can therefore be equated with the class of its potential falsifiers. Accordingly, the more *inclusive* is the class of a statement's potential falsifiers, the more information about the world does the statement provide. Moreover, since Popper believes that the opportunities for falsifying a statement are greater the more inclusive is its empirical content, he maintains that the "risk" of making mistaken assertions also increases with the increasing content of asserted statements.

So much for what are largely matters of terminology; I can now briefly summarize and then examine the main steps in Popper's argument. (1) In the first place, not only are universal statements unverifiable, but according to him this is so for all basic statements as well. "The statement 'Here is a glass of water' [or 'Here is a potentiometer which reads 145']," he declares, "cannot be verified by an observational experience. The reason is that the *universal terms* which occur in [them] ('glass,' 'water' [or 'potentiometer']), are dispositional; they 'denote physical bodies which exhibit a certain *law-like behavior*.' . . . That the instrument is in fact a potentiometer cannot be finally established or verified—no more than that the glass before us contains water. . . ."[7] Indeed, *all* terms are said to be "dispositional" or theoretical, though some are more so than others. "If 'breakable' is dispositional," Popper writes, "so is 'broken.' . . . We should [not call] a glass 'broken' if the pieces would fuse the moment they were put together," for the criterion for being broken is *behavior under certain conditions*. Similarly, "red" is dispositional: "a thing is red . . . if it 'looks red' in certain situations. But even 'looking red' is dispositional. It describes the disposition of a thing to make onlookers agree that it looks red."[8] In short, every basic statement is a hypothesis or conjecture, whose epistemic status does not differ from that of universal statements—since the observations that are commonly (but according to Popper mistakenly) taken to *warrant* or "justify" basic statements, involve *interpretations* of experience in terms of *other theories*. In consequence, although he thinks that the acceptance or rejection of a basic statement may be "motivated" or "caused" by observations, acceptance or rejection is logically a "free decision" which is not, and cannot be, *justified* by experience.[9]

Popper seems to me entirely sound in maintaining that even apparently simple perceptual judgments involve "interpretations" of immediate experience in terms of what he calls "background knowledge" that is accepted as unproblematic in a given context of inquiry—though he does not explicate the sense of "interpretation" that is relevant to his claim, and indeed does not consider how concepts are related to the subject matters encountered in sensory experience, despite the fact that this is a crucial question for the views he advocates. Moreover, I also think he is right in noting that various parts of this background knowledge may nonetheless become problematic in other contexts of inquiry, so that the perceptual judgments which presuppose them are corrigible and may need to be revised—although in this connection also he does not explore the implications of his incidental but by no means insignificant dictum that "almost all of the vast amount of background knowledge which we constantly use in any informal discussion will, *for practical reasons, necessarily remain unquestioned*."[10] However, neither of these claims is likely to appear unfamiliar or dubious to anyone who has read Kant and Peirce with any care. On the other hand, I fail to see why, granted that perceptual judgments are in principle all corrigible, we must conclude that *whenever* we assert a statement such as "Here is a glass of water" or "Here is a potentiometer reading 145" on the basis of observations, we are simply making a conjecture, and cannot rightly hold that we *really know* such statements to be true because in asserting them we allegedly run a genuine risk of being in error—even if the observations on which those statements are grounded have been made under the most carefully controlled conditions. The claim that there *always* is such a risk is hardly supported by *factual* evidence, since in an untold number of cases in which, for example, something has been judged to be a glass of water, there simply are *no* reasons for supposing that an error has been committed, and excellent reasons for supposing the contrary. Mistaken perceptual judgments admittedly do occur, just as there undoubtedly have been convictions of the innocent. But it surely does not follow that because men have *sometimes* been erroneously declared to be dead, no physician is ever *fully justified* in certifying that one of his patients had died; or that since some men have suffered from a miscarriage of justice, a court is never *fully warranted* in holding any accused person to be guilty as charged. Genuine doubt as distinct from purely verbal professions of uncertainty, as Peirce made clear, is based on *positive* reasons; and the sound fallibilistic reminder that every cognitive claim *may* need revision in the light of further inquiry, so that we should be alert to sources of possible error and be prepared to correct them when they occur, is *not* a sufficient ground for regarding every statement as merely a conjecture and for doubting whether we ever have knowledge.

Moreover, consider Popper's contention that in accepting a basic statement ostensibly because of some observation, we are simply stopping "at a kind of statement that is especially easy to test" and on which "various investigators are likely to reach agreement";[11] and also his further claim that the acceptance of a basic statement is a "convention" or "agreement" on "a practical matter"[12]—a "free decision" which may be motivated or caused by our observations, but which cannot be justified by them.[13] But is an agreement among investigators on a basic statement just a coincidence, or do they agree to accept the statement because they agree on what they *observe*? In either case, is the assertion (indeed, the basic statement) that a certain group of investigators agree on a basic statement just another unjustifiable "agreement" on "a practical matter"? If it is, since theories are falsified and hence rejected when they contradict accepted basic statements, why is not the alleged scientific search for true explanations just a game whose various moves and outcomes depend entirely on the arbitrary decisions of so-called "investigators"—and just how, if at all, does the position taken differ from the contention of some sociologists of knowledge that what scientists of a given society and period regard as true depends entirely on their social perspectives? Accordingly, is anything really clarified by saying that the acceptance of a basic statement is a decision taken for practical reasons? Are such decisions really "free" and all equally good? But if not, how else can their comparative worth be evaluated except by reference to matters of observed fact? And what would a "justification" for accepting a basic statement *be,* if no amount or kinds of observation can provide one? Are the only cognitive claims that can be justified those which are about purely formal relations of logical deducibility or opposition between statements? Or is the point simply that the *word* "justification" is not to be used in discussing the grounds for claims to perceptual knowledge? But if so, is the problem eliminated by eliminating the word? These and similar questions that could be raised suggest that a coherent fallibilism, though steadfast in insisting that no empirical statement is *inherently* incorrigible, cannot dispense with the notion that statements about observable matters may differ in their reliability, and it cannot rule out on *a priori* grounds the possibility that some claims to knowledge are actually warranted.

(2) Nevertheless, despite Popper's repeated and explicit claim that no empirical statement can be known to be true or even reliably established, I confess to being sometimes of two minds whether he is

arguing for anything more than a proposal to ban the *word* "knowledge" and its cognates, and to replace them by others (such as "corroborate" or "prove its mettle") more to his taste. Were this the sole question raised by him, it would be silly to pursue it any further. However, much more appears to be involved than a pointless quarrel over terminology in his account of what he thinks is the rational strategy of science, and I will therefore examine that part of his analysis which deals explicitly with scientific theories.

As has already been mentioned, although according to Popper theories cannot be verified, they can be corroborated or confirmed—that is, put to severe tests. Moreover, theories may be corroborated to a greater or lesser extent; and he even suggests several ways in which a numerical measure can be assigned to the degree to which a theory has been corroborated. At first sight, it might therefore be supposed that the locution "degree of corroboration" (or the equivalent phrase "degree of acceptability," which he also uses occasionally[14]), is simply Popper's way of talking about what is often called the degree of rational confidence in the truth of a theory that is warranted or justified by the available evidence for it. But this is not a tenable supposition. For he not only denies this explicitly,[15] he also makes clear that his degree of corroboration "is nothing but a measure of the degree to which a [theory] has been tested,"[16] so that far from being capable of serving as an index for *affirming* a theory, the degree of confirmation shows only how "worthy" the theory is of being selected for further severe testing in view of its having survived "sincere efforts to overthrow" it.[17]

The corroboration of a theory is therefore just a by-product and not the aim of the process of trying to falsify it. Popper's main and indeed almost exclusive stress is upon the virtue of science as a persistent attempt to *refute* theories, rather than to find *supporting evidence* for them: and he adopts a criterion of scientific progress which minimizes the successful *resolution* of the problems that initiate specific inquiries, and gives overwhelming priority to the continuing search for new ones. Thus, he devotes much effort to showing that a theory with a more inclusive empirical content has a smaller logical "probability" than does a theory with a less inclusive one, since the former provides more opportunities than does the latter for being falsified in a wide variety of circumstances. He therefore argues that progress consists in developing "daring," "highly informative," but "improbable" theories,[18] which employ "simple, new and powerful, unifying idea[s] about some connection or relation between hitherto unconnected things."[19] For such theories can be stringently tested; and though the risk that they will be refuted is great, their overthrow is not a defeat for science. On the contrary, the falsification of such a theory "should be regarded as a great success"—a success, because the refutation is a witness to the fertility of the theory to suggest new experiments, and because "science would stagnate, and lose its empirical character" if theories were not refuted. The inability of a theory to stand up under stringent tests is an incentive for developing a more comprehensive theory, which will meet not only the tests that the refuted theory passed as well as those it did not, but will in turn also suggest fresh problems and previously untried tests to which the new theory can be submitted.[20] Accordingly, he declares that "the most lasting contribution . . . that a theory can make are the new problems which it raises."[21] Although Popper maintains that at no point in this process of repeatedly overthrowing theories and replacing them by others do we have *justifiable* grounds for believing that we are approaching "more and more closely to the truth,"[22] he nevertheless calls this process, curiously enough, "the growth of scientific knowledge."[23]

This is a Mephistophelian view of science. It is an account according to which to suppose even for a moment that any problem is really solved is to lose one's soul; and science must say with the Prince of Darkness: *"Ich bin der Geist der stets verneint!"* It evokes in me the nightmarish image of a barbaric ritual, in which the successes of a participant in escaping from attempts on his life, ostensibly undertaken with the intent to test his prowess, are nevertheless never used as a basis for assessing his abilities for performing well under similar circumstances, but are noted merely to single him out as a suitable candidate for even more strenuous efforts to kill him. But however this may be, if Popper's analysis is essentially sound, science appears to be indeed a quest for uncertainty with a vengeance.

How sound is it actually? Science is undoubtedly an enterprise of continuing inquiry into *problems* with varying degrees of generality and related to diverse subject matters. It is an ongoing inquiry, partly because one question commonly leads to another, a high value is frequently placed by those engaged in it upon the raising of new ones, and virgin areas for systematic research are discovered. The various things Popper says on this score, as well as on the critical rationality of science as an honest search for evidence that has eliminative and not merely illustrative force, seem to me admirable. However, though the history of science is the history of human effort to resolve problems, it is not the *same* problem that men continue to discuss. Moreover, although some questions have received either no answers thus far, or admittedly tentative answers, or answers once believed to be correct but subsequently shown to be mistaken, it is surely an exaggeration to say that *no* problems have ever been settled, or that *no* theoretical (that is strictly universal) statements are accepted in science as fully warranted. That a certain liquid substance called water expands its volume when it is under a certain pressure and is cooled to a little above its freezing temperature; that this substance decomposes under certain specifiable conditions into two identifiable gases called hydrogen and oxygen; that glowing hydrogen gas emits radiation characterized by the presence of certain distinctive spectral lines under statable circumstances, are just three examples of physical laws which no physicist today would label as "conjectures" or declare to be anything but well established. These laws were the solutions that were found to certain problems, and they terminated the particular inquiries that were generated by those problems. To be sure, the laws do not give answers to all the questions that have been asked, or continue to be asked, about water or hydrogen. Nor do they necessarily settle the questions to which they *are* answers with the degree of precision that may be required in some investigations—for example, it may not suffice for certain purposes to be told that the temperature at which water begins to expand lies in the interval from 3.9° C to 4.1° C, and new studies may have to be undertaken to deal with the problem whether a smaller value for this interval can be fixed. But the truism that the outcome of an inquiry is not an adequate or relevant answer to *all* problems lends no support to the thesis that *every* answer advanced by inquiry into any *given* problem is at best only a conjecture.

Moreover, although theories are frequently modified and sometimes discarded in the light of new evidence, Popper appears to me to exaggerate the role which refutations play in this process. For he assumes that once a basic statement has been accepted that describes some "reproducible effect" and contradicts a theory, the theory is falsified.[24] This is *formally* correct. However, the relevance of the assumption to any actual inquiry obviously depends on whether the basic statement really contradicts the theory; and this question does not always have a straightforward answer. In the first place, many theories and laws (especially though not exclusively in the natural sciences) formulate relations of dependence between things for what are variously called "pure," "ideal," or "limiting" cases. For example, the familiar laws about the conditions for equilibrium of simple levers are stated for levers that are weightless but perfectly rigid and straight rods. Accordingly, since actual bodies are at best only "approximations" to the ideal requirements of laws so formulated, a basic statement apparently incompatible with such a law can often be properly regarded not as *refuting* the laws, but as indicating an area in which the law as stated is simply not *applicable*.

Second, and irrespective of whether laws are formulated in terms of pure cases, although a theory may be accepted because it has survived critical tests in some given area, no *unambiguous commitment* might be made concerning the *precise* range of phenomena for which the theory is accepted. In consequence, if a basic statement is accepted in apparent conflict with the theory when the theory is applied *outside* the initial area, the conflict could be resolved in several ways without rejecting the theory as falsified. Thus, on the assumption that the theory continues to survive testing in the initial area, its scope could be explicitly restricted to that area. Or some factor or condition not mentioned in the theory might be found, variation in which differentiates the area where the theory works well from domains where it does not; and it may be possible to exhibit the theory as a special case of a more exclusive one. For example, Boyle's law holds for some gases at certain constant temperatures, though not at others, the difference being associated with variations in certain identifiable factors not stated in the law; but instead of rejecting it as falsified by observations on gases at certain temperatures, Boyle's law is frequently accepted as a limiting form of another law which *does* mention those factors.

And, third, Popper's account of falsification is predicated on the assumption that every empirical theory is associated with a fixed and unambiguously definable class of potential falsifiers. Now it is not difficult to specify such a set of basic statements for theories of a relatively simple sort—to cite one of his own examples, for such strictly universal statements as "All ravens are black." But it is far from easy to do this, and perhaps even generally impossible, for the "daring," "highly informative," and "unifying" theories (for example, Newtonian gravitational theory or electromagnetic field theory) that are so conspicuous in some of the natural sciences, and whose role in the growth of knowledge Popper prizes so highly.

The difficulty in defining the potential falsifiers for theories of this sort derives from several sources, but I will limit myself to mentioning just one. These theories are not single statements, but *systems* of assumptions, whose component statements contain terms, such as "atom" or "light ray," which for the most part do not designate anything identifiable either by direct observation or even by laboratory instruments. Accordingly, if such theories are to be tested experimentally, they must be supplemented by a distinctive set of further statements—variously called "coordinating definitions," "correspondence rules," and "bridge laws"—in order to establish connections between the theoretical entities postulated by such a theory and certain matters that can in fact be observed. However, bridge laws are rarely formulated precisely or even explicitly; but what is more important, the points at which connections between theory and experience are introduced can be varied according to the particular research uses that are made of the theory, so that the number of bridge laws (and hence the experimentally confirmable consequences of the theory) does not remain constant, though without thereby changing the internal structure and meaning of the theory. In consequence, it is extremely difficult to decide what is to count as a potential falsifier for such a comprehensive theory, since the introduction of new bridge laws or the elimination of old ones may shift the status of an accepted basic statement from that of being a falsifier to that of being compatible with the theory, or vice versa. For example, whether a given measurement on the angles of a given configuration is to count as a refutation for some geometric theory will depend on what coordinating definitions have been adopted to relate the theory to experimental material.

Moreover, the system of theoretical statements is usually quite flexible, since it may allow for various modifications within the framework of certain broad assumptions, though without explicit restrictions on the possible modifications within those assumptions. Thus, Newtonian mechanics postulates forces to account for accelerations, but does not prescribe a particular form of the force function that may be used in analyzing the motions of bodies, and thereby grants considerable leeway for the introduction of various types of forces; nor does Newtonian mechanics exclude the use of statistical hypotheses in the study of the motions of a large number of bodies, where the statistical hypotheses may themselves take different forms. Accordingly, as Duhem noted, an apparent conflict between such theories and observation may be resolved by introducing a suitable modification in the theory, so that a basic statement that was initially taken to be a falsifier of the theory loses that status.

Indeed, what is sometimes called a theory is often a *family* of more or less similar theories; and though a particular member of the family may be falsified, this does not overthrow the indefinitely large number of other members of the family. Viewed in this perspective, the claim that Newtonian mechanics, for example, has been refuted makes little sense—if "refutation" is understood in the manner that Popper prescribes for the word. To be sure, certain familiar forms of Newtonian theory—that is, certain members of the Newtonian family of theories—have been found to be inadequate in several fields of physical inquiry; and though it is most unlikely that some still untried forms of Newtonian theory will replace the currently employed theories in those areas, this remains an open logical possibility. It is not the simple falsification of some particular form of a comprehensive theory that completely decides the theory's fate.

My comments on Popper have assumed that his account of the logic of science is in the main a reflection on the *actual* procedures of science, and that his analysis can therefore be evaluated in the light of one's familiarity with those procedures. On such an evaluation, I believe he is correct in noting the fallibilism that is inherent in science as an *ongoing* and *institutionalized* inquiry, irrespective of the conduct of individual scientists or of their views on this question, as well as in pointing out the indispensable role of theories in the conduct of research and the growth of science. On the other hand, I have tried to argue that Popper is not a dependable guide on the question whether warranted claims to knowledge can be established through critical inquiry, and that his conception of the role of falsification in the use and development of theories is an oversimplification that is close to being a caricature of scientific procedure.

However, much that Popper says suggests that his intent is not to offer a generalized *description* of the logic in use in science, but on the contrary to argue for a *prescription* he is recommending on how science *ought* to proceed. If his analyses are construed in this way, his prescriptions seem to me to be based—though I have not the time to develop the point in this paper—mainly on *formal considerations,* and to presuppose a conception of what is a warranted claim to knowledge that is standard in the *formal disciplines* such as pure logic or demonstrative geometry. In a formal science, the "evidence" for all statements asserted in it must have demonstrative force—i.e., every such statement must follow logically from the evidence for it. However, since no empirical statement can be asserted on this ground, no empirical statement—so Popper appears to say—*should* be asserted, whatever the available evidence. But we come close to demonstration when we show that an empirical statement *contradicts* some conventionally adopted basic statement; accordingly, so he seems to recommend, empirical science *should* seek to refute theories, rather than to find supporting evidence that might warrant rational belief in them, since the evidential statements neither will be demonstrably true nor can they imply any theory with demonstrative force. I need hardly say that this argument as just stated is my reconstruction of Popper's grounds for his policy recommendations, though I do not think I have distorted his views beyond recognition. But in any case, I do not find the substance of his ideas on science, when understood to be prescriptive, any less dubious than when they are taken to be descriptive.

2

Popper's general views on the logic of scientific inquiry are shared by a number of contemporary thinkers, some of whom have arrived at their conceptions independently of his. In the remainder of this paper I want to examine some implications that have been drawn from the version of fallibilism those views represent, especially since the conclusions drawn have been used to analyze several central issues in the philosophy of science in a more explicit and detailed manner than Popper's own writings exhibit. However, since my available time is short, I must restrict myself to discussing briefly the account of scientific explanation recently advanced by Paul Feyerabend.

Like Popper, to whom he makes numerous acknowledgements, Feyerabend maintains that science is a search for true explanations, but that we can never rightly claim to know the truth. However, although his argument for the latter contention is the familiar one Popper uses, he carries it to more radical extremes, and I will therefore set it out in two stages. (1) In the first place, all so-called "observation statements" occurring in the natural sciences are held to be "theory-impregnated" in a double sense: in the sense that they assert more than is ever "given" in experience, so that they can never be completely verified but have the status of hypotheses; and in the sense that the *meanings* of the terms in them can be explicated only by way of theoretical assumptions, so that changes in theory bring with them changes in the meanings of all observation terms, and therefore changes in the commitments to what Feyerabend calls the "ontologies" postulated by theories.

Since this notion of "changes in meaning" is central to Feyerabend's views on scientific explanation, it requires further exposition. He supports his claim that the meanings of observation terms are not invariant under changes of theory by a discussion of instrumental measurement, noting that:

> . . . the indications of instruments do not mean anything unless we possess a *theory* which teaches us what situations we are to expect in the world, and which guarantees that there exists a reliable correlation between the indications of the instrument and such a particular situation. If a certain theory is replaced by a different theory with a different ontology, then we may have to revise the interpretation of *all* our measurements. . . . Nobody would dream of demanding that the meanings of observation statements as *obtained with the help of measuring instruments* remain invariant with respect to the change and progress of knowledge. Yet, precisely this is done when the measuring instrument is a human being, and the indication is the behavior of this human being, or the sensation he has, at a particular time.[25]

An example will help make clear what Feyerabend is asserting. We are patently assuming certain laws of electricity and magnetism when, upon observing the position of the pointer on the scale of a properly connected ammeter, we describe the state of the wire in the circuit by saying that it is carrying a current of ten amperes. Were we to reject those laws and replace them by others, but continue to use the same sentence to describe the state of the wire—though it is possible that with the change in laws the sentence no longer has any use—the meaning of the sentence would presumably also have changed.

I will ignore what may be only linguistic slips in his statement, and comment on its essentials. In terms of the example, part of what he says is obviously sound, but this part hardly establishes his claim that change in theories carries with it changes in the meanings of *all* observation terms. It is true enough that if our present laws relating electric current and magnetic field were for some reason discarded, terms such as "ampere" and "current," into whose definitions these laws enter, would cease to have their present use. However, if we continue to use the instruments we call ammeters, and made observation statements like the above, we could make them only because we would still be able to make such observations as "This is a piece of wire," "This pointer moved from one position on the scale to another," or "A connection was closed before the pointer began to move." But there is no reason whatsoever to suppose that the meanings of these statements underwent a change as a consequence of the indicated change in the laws; nor is there reason to believe that their meanings would change if further progress were made in other branches of knowledge.

Moreover, though descriptive terms (such as "current" or "ampere") that are defined by way of some theory are often employed to describe what are commonly alleged to be observable phenomena, such use of terms is subject to important limitations. For example, suppose that some theory is being tested by obtaining certain experimental data. It would be clearly circular to interpret those data in such a way that they are described with the help of terms whose definitions presuppose the truth of the theory. Accordingly, observation statements containing such terms can be used in this case, only on pain of making the theory irrefutable, and thereby depriving it of empirical content. There must therefore be observation terms whose meanings are neutral with respect to the theory being tested, and indeed whose meanings are invariant under changes in accepted theories of a certain class. Although the point cannot be developed here, it seems to me that in advancing his radical thesis concerning the meaning of observational terms, Feyerabend is confounding the meaning of a descriptive term, in the sense of "meaning" which equates meaning with what is logically entailed by the *connotation* (or definition) of a term, with the meaning of a term in the sense of "meaning" which equates meaning with what is logically entailed by the *empirical laws* into which the term enters.

(2) But I must turn to the second part of his argument, which deals with scientific explanation. The main thesis he seeks to establish in this connection is that when a law or theory (for example, Galileo's law of free fall, or the three Keplerian planetary laws) is explained by a more comprehensive theory (for example, the Newtonian theory of gravitation), its "ontology" is completely replaced by the ontology of the more comprehensive theory, so that the meanings of all its descriptive terms are changed. He therefore rejects the widespread view that a necessary condition for the explanation of one theory by another is that the first be formally deducible from the second, so that the first is incorporated without alteration into the second. Accordingly, he portrays the progress of science as the creation of increasingly comprehensive but *incommensurable* theories, with corresponding transformations in the meanings of the observation terms employed to describe the data that serve to test the theories. As long as theories continue to change, there can therefore be no continuing body of stable knowledge that is transmitted from one generation to another; and indeed, the notion of progress in knowledge remains something of a mystery on this analysis.

But I must examine the grounds on which Feyerabend denies that explanations ever take a deductive form. It will be convenient to distinguish between two *prima facie* different types of explanations: those in which the law or theory to be explained contains no terms essentially different from the terms contained in the more comprehensive theory, and those in which it does contain such terms and which are sometimes called "reductive" explanations.

(a) Examples of the first type discussed by Feyerabend are the familiar explanations of Galileo's law and the Keplerian laws by Newtonian theory, all of which are frequently mentioned as illustrations of deductive explanations. But he argues that this is really not the case. For according to Galileo's law, the acceleration of a freely falling body is constant; however, according to the law that can be deduced from Newtonian assumptions, the acceleration is not constant, since it is inversely proportional to the square of the falling body's distance from the earth's center and therefore varies as the body falls. Even though for short distances the difference between the Galilean and

Newtonian propositions is experimentally indistinguishable, Feyerabend claims that a *conceptual* difference remains, so that Galileo's law cannot strictly be deduced from Newtonian theory. Accordingly, if Newtonian theory is adopted, Galileo's law must be rejected as *false* and *replaced* by another law which *contradicts* but at the same time corrects it. A similar argument is used in connection with Kepler's laws, but I will not go into it, since all the essential points are made in simpler form in connection with the Galileo example.

Let me note in passing that Feyerabend himself recognizes an explanation that is deductive—namely, the explanation of the law which is deduced from Newtonian theory. But there are two more important points on which I want to make brief comment: (i) Strictly speaking, Feyerabend is quite right in claiming that Galileo's law cannot be deduced from Newtonian assumptions *as he states them*. However, if one inspects what are called the "deductions" of other laws from Newtonian theory as these are presented in books on mechanics—for example, the deduction of the law according to which the period of a simple pendulum is proportional to the square root of its length l divided by g (the "constant" of acceleration)—it becomes evident that the derivations frequently introduce some simplifying assumptions into the premises. In the case of the pendulum law, the assumptions are that the angle through which the pendulum oscillates is quite small, and that the acceleration is constant; and an analogous assumption is made in the customary derivations of Galileo's law. Accordingly, if the explanation of the pendulum law can count as deductive in form, so can the explanation of Galileo's law; and appearances to the contrary in the case of Galileo's law stem from the fact that Feyerabend does not make explicit all the premises required for its deduction. I am of course not denying that something new is added to Galileo's law when it is deduced from Newtonian theory; for the deduction makes clear some of the *limits* within which the law holds, and indicates how it needs to be *corrected* outside those limits. But to say this is something quite different from saying that Galileo's law must be rejected as false and replaced by a law conceptually different from it.

(ii) Furthermore, Feyerabend completely overlooks the important point that Newtonian theory is formulated in what I have previously called "ideal" terms (such as "instantaneous acceleration"), so that it cannot yield any experimentally significant consequences unless, among other things, various coordinating definitions or correspondence rules are introduced. But such "bridge statements" presuppose that the descriptive observation terms contained in them have relatively stable meanings which are invariant in the process of explanation. In short, theoretical science would be impossible if everything were as conjectural and uncertain as Feyerabend's thesis maintains.

(b) The example of the reductive or second type of explanation Feyerabend discusses is the explanation of classical (or phenomenological) thermodynamics in terms of statistical mechanics, a special case of which is the explanation of the familiar Boyle-Charles gas law ($pV = kT$) by the molecular theory of gases. Even this special case of the general problem is far too technical to permit my examining the issues with adequate detail in this paper, and I can mention only what I think is the central point at stake. Since the Boyle-Charles law contains the term "temperature," but the molecular theory does not, the question is: How is it possible to explain the former by the latter, if the classically defined notion of temperature makes no reference to such things as molecules that are postulated by the molecular theory? Feyerabend's answer is that the explanation is not deductive in form; and he rejects the view that it does take this form, even if a bridge law is introduced which connects the thermodynamical notion of temperature with the notion of the average kinetic energy of molecules as defined in molecular theory. According to him, on the contrary, the explanation consists in *replacing* the Boyle-Charles law by another, in which the term "temperature" must be explicated by way of the ontology of molecular theory, and no longer means what it meant prior to the explanation. He also maintains that if the molecular theory is accepted, we must reject as false not only the Boyle-Charles law, but also the second law of thermodynamics in its classical formulation.

But it is difficult to understand why, if thermodynamics is as false as Feyerabend says it is, physicists continue to study the subject and to use the theory in many of their investigations, even after they have mastered the kinetic theory of matter. They certainly exhibit no obvious signs that they are dealing with incommensurable concepts when they pass from researches employing classical thermodynamics to inquiries using statistical mechanics. Moreover, it is far from clear that the second law of thermodynamics has been shown to be false by investigations on Brownian motion, as Feyerabend alleges, since in such motions the "entropy" of the system sometimes decreases. For as Max Planck pointed out, the classical notions of temperature and entropy cannot be *applied* to such systems, because the defining conditions for their application are not realized;[26] and accordingly, what the kinetic theory of matter shows is not that classical thermodynamics is *false*, but rather the *limits* within which it is *true*.

In any event, there is a clear alternative to a view like Feyerabend's, which sees the history of scientific inquiry as a discontinuous progression of incommensurable, unjustified, and unjustifiable conjectures. Santayana contrasted myth and science by noting:

> A geographer in China and one in Babylonia may at first make wholly unlike maps; but in time both will take note of the Himalayas, and the side each approaches will slope up to the very crest approached by the other. . . . So science is self-confirming, and its most disparate branches are mutually illuminating; while in the realm of myth . . . there can be nothing but mutual repulsion and incapacity to understand.[27]

Although I think with Santayana that myth and science differ, it is well to remember that much that passes as science is indeed myth; and the strength of fallibilism lies in the recognition that continuing effort is required not to confound the two. However, the versions of fallibilism I have been considering cannot in the end admit that myth is not the same as science and can often be told apart; and I have tried to show that a fallibilism which leads to such a conclusion is neither coherent nor supported by an adequate analysis of the logic of science.

5/ Philosophical Depreciations of Scientific Method

Obscurantist creeds and superstitious beliefs have, like the poor, been with us for a long time, and it is plain that we have not learned how to eradicate them from our midst. What is especially disappointing about their prevalence and the disdain of science often accompanying them, however, is that they continue to flourish even in societies in which education through secondary school is mandatory and illiteracy has been eliminated. It was the confident hope of the men and women who were in the forefront of the struggle to make such an educational opportunity available to all, that if this goal were realized at least the more blatant superstitions would disappear. In fact, however, there are communities in the United States in which programs of adult education sponsored by local high schools include courses on astrology and witchcraft; and a large assortment of other pseudoscientific beliefs continue to be widely held. Whatever may be the reasons why those early expectations of what mass education would bring have not been realized, it is evident that the task of developing throughout society a critical attitude toward ideas is much more difficult than is commonly supposed.

Explicit attacks on science as an institutionalized activity for obtaining reliable knowledge occur in many forms. Scientific accounts of nature have been criticized as irremediably inadequate, because they explain the behavior of things in terms of allegedly ultimate components that possess none of the inexhaustibly rich qualities displayed in our daily encounters with the world, so that those accounts—especially in the physical sciences—describe nature as nothing but a congery of abstract relational structures. Alternatively, scientific theories are said to have no standing as descriptions of the furniture and the mechanisms of nature, and to have only the instrumental value of enabling men to order their experiences in a convenient manner. Again, various principles employed in evaluating the conduct and outcome of scientific inquiry have been characterized as dogmatically imposed restrictions on what can properly count as a scientific investigation; and in consequence, those principles are also said to exclude on *a priori* grounds views of nature that fail to meet arbitrarily selected criteria of what constitutes authentic knowledge.

Critics of the view that scientific inquiry is the most effective way of obtaining warranted knowledge are found not only among those who are for the most part strangers to pure and applied science. Indeed, perhaps the most serious attacks on that view have come from within the ranks of philosophers and historians of science--from men whose accounts of scientific procedures and interpretations of scientific findings in effect deny both that a scientific theory is ever adopted in the light of a rational evaluation of the evidence, as well as that the validity of cognitive claims in the sciences is invariant despite differences in social and personal outlook. In my opinion, these denials of the very possibility of objectivity in scientific inquiry are a more serious threat to the continuance and spread of a scientific attitude in resolving intellectual problems than is the growth in the audience for astrology and other superstitions. For an increase in the number of believers in *specific* superstitions may leave unaffected a *widespread* acceptance of the authority of scientific method in all but a few limited domains. On the other hand, those attacks on the notion that scientific inquiry can be objective are tantamount to an endorsement of the view that the grounds on which conclusions in the sciences are accepted are at bottom no better than are the grounds on which superstitious beliefs are adopted. Those attacks may therefore undermine general confidence in the superiority of scientific method over other ways of validating beliefs, and so justify almost any doctrine, however unwarranted it may be, on the ground that the dubious considerations which may be invoked in its support are not inferior to those on which conclusions in the sciences are based. I want therefore to examine briefly three related claims, whose general import is to deny that scientific inquiry ever achieves genuine knowledge or employs determinate rules for evaluating the cognitive worth of its intellectual products.

1

The first of these claims deals with scientific method, and advocates the thesis that the growth of knowledge is best served by rejecting definite rules for the conduct of inquiry, and adopting instead an "anarchistic theory of knowledge." According to this thesis there not only *are* no strictly universal rules which practicing scientists employ for determining whether a proposed hypothesis is to be accepted or rejected; the thesis also asserts that it would be *undesirable* to have such rules. "The idea of a method," a vigorous proponent of this thesis declares, "that contains firm, unchanging, and absolutely binding principles for conducting the business of science meets considerable difficulty when confronted with the results of historical research. We find, then, that there is not a single rule, however plausible, and however firmly grounded in epistemology, that is not violated at some time or other. . . . Such violations are not accidental events. . . . On the contrary, . . . they are necessary for progress." For example, the Copernican Revolution and the emergence of the wave theory of light are said to have occurred only because "some thinkers either *decided* not to be bound by certain 'obvious' methodological rules, or because they *unwittingly broke* them."[1] Indeed, so it is alleged, for any given rule (such as the rule that no hypothesis should be introduced which violates either well-confirmed theories or well-established experimental results) there always are circumstances when it is advisable not only to ignore the rule, but to adopt its opposite. The conclusion drawn from all this is that "there is only *one* principle that can be defended under *all* circumstances and in *all* stages of human development. It is the principle: *anything goes.*"[2]

However, the argument for methodological anarchism does not support the thesis that anything goes in scientific inquiry. a) In the first place, it is a straw man that is being demolished, not a reasonably adequate conception of method in inquiry, when criticism is leveled against the notion of method that subscribes to "firm, unchanging, and absolutely binding principles for conducting the business of science"—against principles stated without qualification and without reference to the contexts in which the principles are to be used. It would indeed be absurd to adopt a rule to the effect that no theory is to be proposed which is incompatible with well-confirmed theories, *if* the rule is to be followed in *all* contexts—for example, in the circumstance in which the theory, though well-confirmed in a large class of phenomena, seems nevertheless unable to account for some relevant but also basic phenomenon. On the other hand, the absurdity would be absent if the rule were formulated to the effect that no theory should be proposed that is incompatible with well-confirmed theories, unless these latter fail to account for some phenomenon falling into domains in which those theories have otherwise been successful. For example, one consideration that may have motivated the creation of the general theory of relativity is that, though the Newtonian theory of gravitation was well-confirmed in dealing with most planetary motions, the latter appeared unable to give an adequate account of the motion of the perihelion of the planet Mercury.

b) In the second place, there is a blatant nonsequitur in arguing that because any given methodology has limits,[3] one should abandon all methodologies. Thus, every cutting tool (e.g., a chisel, a plane, a saw, a rasp, sandpaper, etc.) has limits of what it can do effectively. But it surely would be folly to conclude from this that cutting tools should never be used, or that there cannot be sound and useful rules for using them. And similarly for methodological rules: although every methodology may have limits (e.g., the methodology for deductive inference is not applicable to nondemonstrative reasoning), there may be, and in fact are, determinate rules for employing various procedures in specified contexts.

Moreover, the fact that any conceivable methodological principle cannot be reasonably applied in every possible situation, surely does not warrant the conclusion that "anything goes" in the conduct of scientific inquiry. For example, even if one agrees that it would be unreasonable to subscribe to the methodological rule not to propose hypotheses inconsistent with well-confirmed theories, it does not follow that it would be *reasonable* in such cases to ignore the available evidence for the hypothesis that has been introduced, or to accept that hypothesis without examining the consequences logically implied by it. To be sure, there is no *a priori* formula for determining what strategies of research are in the long run the effective ones, so that there is a sense in which anything goes—namely, the sense that *prior to employing them* in conducting inquiry, all methodological rules are *candidates* for adoption, and that only experience in applying a rule can provide the needed evidence for deciding whether or not the rule contributes to the success of inquiry.

2

I must now examine the second of the three claims which (in effect, though for the most part unwittingly) help to undermine confidence in scientific inquiry as a source of genuine knowledge. This second claim deals with the nature of the support that conclusions of inquiry receive from empirical evidence. There is time for only two brief comments on this question.

a) According to a widely held view, sensory observation may *suggest* what singular statements are to be asserted concerning those observations, but does not *demonstrate* the truth of those statements. In consequence, no statement about matters of ostensibly observable fact can be certified as *true*, so that such singular statements are included in the corpus of scientific knowledge only in virtue of *conventions* that have been implictly made about those statements. Similarly, it is also frequently maintained that since general statements formulating the laws and theories of the various sciences are not entailed logically by any finite class of singular statements, no law or theory of any empirical science can rightly be said to be *known* to be true. In short, the claim is that no statement with an empirical content can be known to be true, so that irrespective of the quantity and nature of the supporting evidence, all statements of the sciences purporting to have a factual content are always subject to revision and are never more than "conjectures." But acceptance of this claim leads easily to a pervasive scepticism concerning the competence of inquiry to yield genuine knowledge.

However, the scepticism is not warranted, and is generated largely because the application of the word "knowledge" is restricted in an arbitrary manner. For example, it is quite correct to say that observation does not establish *demonstratively* the truth of any singular statement about what is being observed. But this is so because in the strict sense of the word "demonstratively," a necessary condition for demonstrating the truth of a statement is that it be shown to follow logically from a set of *premises*—a condition that by assumption is not satisfied by statements that are asserted on the basis of sensory observation. Moreover, it is also the case that singular observation statements as normally used (for example, the statement "That bird is a swallow") assert by implication more than what is *actually* observed (for example, that the object is an animal that was hatched from an egg), so that such statements are subject to revision; and it is evident that any finite number of instantial premises that are supposedly about what in fact *is* observed, will not logically entail such singular statements. Nevertheless, it is seriously misleading to conclude from all this that no singular observation statement can be *known* to be true. For this conclusion assumes that the only proper use of the word "knowledge" is in connection with statements that have been established demonstratively—a usage that is certainly appropriate in formal disciplines like logic and mathematics but, as the history of science makes plain, in none of the empirical sciences. It is, however, quite arbitrary to restrict the application of the word "knowledge" in this way; and it is only because such a restriction is tacitly made that the paradoxical conclusion is advanced that no matter how much observational or experimental evidence is available for a singular statement of fact, the statement is never more than a guess or conjecture.

It is also worth noting that although every factual statement, whether it is singular or general in form, is in principle always corrigible, most singular statements and many general ones that have been asserted about matters of fact have not needed revision. Indeed, a number of the theses asserted by those who deny that we can ever rightly claim to have genuine knowledge of empirical subject matters presuppose that some factual statements are known to be true. For example, the contention that singular statements are included in the corpus of scientific knowledge by convention is a factual statement about the behavior of men, and cannot be regarded as true by convention on pain of an infinite regress.

b) I come to another sort of deprecating view of scientific method that is based on a consideration of the relation between scientific hypotheses and the evidence for them. A number of influential philosophers of science have argued that there is no such thing as nondemonstrative inference; and according to them, however well a theory is corroborated by observational data, there is no basis for the common supposition that such a theory is a more reliable premise for making prediction than is a less well corroborated theory. Moreover, other thinkers, without necessarily subscribing to that extreme view, have come to doubt the "objectivity" of judgments concerning the "strength" or "weight" of given evidence for a given theory. A frequently mentioned reason for this doubt is that no one has yet succeeded in formulating rules to the satisfaction of the scientific community for assigning numerical measures to the merits of nondemonstrative inferences (i.e., to the degree of support that given evidence provides to a given hypothesis), or for judging the comparative merits of competing hypotheses in the light of the available evidence for them. In the absence of such explicitly stated rules, many philosophers have come to believe that the assessment of the evidential support for a theory will vary with the temperament and the preconceptions of the person making the evaluation, and is inevitably a "subjective" matter. In consequence, so it has been claimed, we must abandon the notion that the "likelihood" of a hypothesis can be established in the light of the evidence for it in a noncontroversial manner—that is, in a manner that is free from the prejudices stemming from personal preference and historical circumstance.

There is time for just two brief comments on these claims:

i) It is beyond serious dispute that just about every mature human being accepts, rejects, and acts upon a large variety of statements, even though the statements are not logically implied or contradicted by the evidence for them. It is also not really doubtful that most people *recognize differences in "weight"* of given evidence for a given statement. For example, the assertion that a certain individual is guilty of the offense with which he is charged may be supported by *no* evidence, by evidence that is *negligible,* by evidence that is *not negligible but inconclusive,* or by evidence that is *compelling though without demonstrative force.* But if this much is admitted, nothing of importance is at stake when the question is discussed whether or not such cases should be *called* instances of nondemonstrative reasoning. What *is* important is whether the evidence for statements can be graded and classified according to its probative force; and it is of little moment what *labels* are to be used when dealing with such situations.

ii) It is again beyond dispute that except in certain somewhat specialized circumstances, there are no rules on which there is good agreement for evaluating quantitatively the "strength" of a nondemonstrative inference, however desirable it would be to be able to do so in an unambiguous manner. It should be noted, however, that if there were such rules, a degree of precision would be introduced which in many cases in not needed; moreover, our present inability to quantify nondemonstrative inferences does not necessarily signify the total absence of "objectivity" in evaluating them. Thus, it is frequently sufficient (e.g., in assessing the evidence for an accusation in a court of law, or for an engineer's estimate of the life of a bridge) to be able to decide whether or not the available evidence for a hypothesis is strong enough to make the risk of error tolerable, should the hypothesis be accepted as a basis for some action; and finer gradations of the evidence may serve no useful purpose. Furthermore, even though it may not be possible to assign a numerical measure for the weight of the evidence for a given statement, it is often possible to evaluate the evidence *objectively*—for example, to judge the evidence to be inadequate, on the ground that the items making up the available evidence are *few* in number, that they are unrepresentative because they do not come from a wide *variety* of relevant domains, that they confirm the hypothesis only with considerable *imprecision,* and so on. To be sure, individuals often differ widely in their assessments of evidence. It is, however, also the case that even when individuals make their assessments independently of one another, they concur in their evaluations more frequently than is compatible with the supposition that evaluations are wholly subjective and idiosyncratic.

3

I turn finally to the last of the three claims which generate doubt about the methods of science. It is today a commonplace that scientific theories (such as classical mechanics or Mendelian genetics) are not derivable from the facts of observation, but are, in Einstein's often quoted words, "free inventions of the human mind." However, as Einstein and many others also emphasized, no scientific theory is self-certifying or self-evidently true; and if a theory is to be judged as a true account of some aspect of nature, it must survive the test of experience—that is, the statements logically implied by the theory must agree with the findings of experiment and sensory observation. It is clear that such a process of testing a theory involves the assumption that the observable facts of nature constitute a final court of appeal which decides whether or not the theory under consideration is tenable; and it is also clear that the alleged facts of observation must therefore be established *independently* of the theory, in a manner that is not *antecedently biased* for or against the theory.

However, this assumption has been recently challenged by a number of critics. They claim that the familiar and widely held distinction between so-called "observation" and "theoretical" terms is untenable since, according to them, *all* observation terms involve some theory and are therefore unavoidably "theory laden." As one writer has put the matter, "the meaning of every term we use depends upon the theoretical context in which it occurs. Words do not 'mean' something in isolation; they obtain their meanings by being part of a theoretical system."[4] Indeed, not only is every observation said to be theory laden; the more radical claim is also made that "each theory will possess its own experience, and there will be no overlap between these experiences." Every theory has its "own" observation terms, whose meanings therefore differ from the meanings of the observation terms associated with a different theory, so that theories which are not logically equivalent "will not share a single observation statement."[5] The net outcome of these dicta is that different theories are also "incommensurable." In consequence, a decision between what would ordinarily be regarded as different but "competing" theories cannot be made on rational grounds—for example, by appealing to *independently* ascertained facts of observation. The decision must rest entirely on "aesthetic judgments, judgments of taste, metaphysical prejudice, religious desires, in short, . . . [on] our subjective wishes."[6]

If these claims were sound, there would be no reason for believing that scientific inquiry is a better way of obtaining reliable knowledge than are the approaches just listed. But are the claims sound? I can indicate only in rough outline why I think they are not: a) There can be no serious doubt that the common distinction between observation and theoretical terms is at best a loose one, and that there are no carefully formulated criteria for membership in either of these classes. It must also be acknowledged, and has already been noted above, that what are called "observation" statements in the sciences as well as in our daily pursuits assert far more than what is actually presented in a momentary experience. There are therefore solid grounds for the claim that observation terms and statements are theory laden. On the other hand, there is no evidence to show or reason to believe that when a theory is being tested experimentally, the theory upon which the meanings of some set of observation terms depend is *identical* with the theory that is *being tested*. On the contrary, most if not all the terms employed in describing the observations that are made with the intent of testing a given theory usually have established meanings that are not assigned to those terms by that *very same* theory. For example, terms like "volume," "pressure," and "temperature," used in recording observations on the behavior of gases, have meanings which are not being defined or otherwise specified by the kinetic theory of gases, when that theory is being tested by those observations.

b) There are also good reasons for doubting the claim that even when two different theories deal with the same phenomenon (such as tidal behavior), they share no terms in common and are incommensurable. In the first place, it is difficult to understand how two theories can be mutually incompatible (e.g., Newtonian gravitation theory and the general theory of relativity, which are often said to be so related), if the meaning of every term in one differs from the meaning of every term in the other. Thus, if two statements are contraries of one another (e.g., the statements "The distance from New York to Buffalo is at least 375 miles" and "The distance from New York to Buffalo is 300 miles"), they must have terms in common; and similarly for two theories.

Secondly, it is simply not true that every theory has its own observation terms, none of which is also an observation term belonging to any other theory. For example, at least some of the terms employed in recording the observations that may be made to test Newton's corpuscular theory of light (such terms as "prism," "color" and "shadow"), underwent no recognizable changes in meaning when they came to be used to describe observations made in testing Fresnel's wave theory of light. But if this is so, the observation statements used to test a theory are not necessarily biased antecedently in favor of or against the theory; and in consequence, a decision between two competing theories need not express only our "subjective wishes," but may be made in the light of the available *evidence*.

I will conclude with a summary. Scientific theories have been frequently likened to myths, and there undoubtly are resemblences between them. However, if the preceding examination of some recent critiques of scientific method is essentially correct, there also are fundamental differences between theories and myths. For that examination shows that the various claims on which those critiques rest are unsound, and that the sceptical relativism in which the claims culminate is unfounded.

6/ Issues in the Logic of Reductive Explanations

A recurrent theme in the long history of philosophical reflection on science is the contrast—voiced in many ways by poets and scientists as well as philosophers—between the characteristics commonly attributed to things on the basis of everyday enounters with them, and the accounts of those things given by scientific theories that formulate some ostensibly pervasive executive order of nature. This was voiced as early as Democritus, when he declared that while things are customarily said to be sweet or bitter, warm or cold, of one color rather than another, in truth there are only the atoms and the void. The same contrast was implicit in Galileo's distinction, widely accepted by subsequent thinkers, between the primary and secondary qualities of bodies. It was dramatically stated by Sir Arthur Eddington in terms of currently held ideas in physics, when he asked which of the two tables at which he was seated was "really there"—the solid, substantial table of familiar experience, or the insubstantial scientific table which is composed of speeding electric charges and is therefore mostly "emptiness."

Formulations of the contrast vary and have different overtones. In some cases, as in the examples I have cited, the contrast is associated with a distinction between what is allegedly only "appearance" and what is "reality"; and there have been thinkers who have denied that so-called "common-sense" deals with ultimate reality, just as there have been thinkers who have denied that the statements of theoretical science do so. However, a wholesale distinction between appearance and reality has never been clearly drawn, especially since these terms have been so frequently used to single out matters that happen to be regarded as important or valuable; nor have the historical controversies over what is to count as real and what as appearance thrown much light on how scientific theories are related to the familiar materials that are usually the points of departure for scientific inquiry. In any case, the contrast between the more familiar and manifest traits of things and those which scientific theory attributes to them need not be, and often is not, associated with the distinction between the real and the apparent; and in point of fact, most current philosophies of science, which in one way or another occupy themselves with this contrast, make little if any use of that distinction in their analyses.

But despite important differences in the ways in which the contrast has been formulated, I believe they share a common feature and can be construed as being addressed to a common problem. They express the recognition that certain relations of dependence between one set of distinctive traits of a given subject matter are allegedly explained by, and in some sense "reduced" to, assumptions concerning more inclusive relations of dependence between traits or processes not distinctive of (or unique to) that subject matter. They implicitly raise the question of what, in fact, is the logical structure of such reductive explanations—whether they differ from other sorts of scientific explanation, what is achieved by reductions, and under what conditions they are feasible. These questions are important for the understanding of modern science, for its development is marked by strong reductive tendencies, some of whose outstanding achievements are often counted as examples of reduction. For example, as a consequence of this reductive process, the theory of heat is commonly said to be but a branch of Newtonian mechanics, physical optics a branch of electromagnetic theory, and chemical laws a branch of quantum mechanics. Moreover, many biological processes have been given physicochemical explanations, and there is a continuing debate as to the possibility of giving such explanations for the entire domain of biological phenomena. There have been repeated though still unsuccessful attempts to exhibit various patterns of men's social behavior as examples of psychological laws.

It is with some of the issues that have emerged in proposed analyses of reductive explanations that this paper is concerned. I will first set out in broad outlines what I believe is the general structure of such explanations; then examine some difficulties that have recently been raised against this account; and finally discuss some recent arguments that have been advanced for the view that a physico-chemical explanation of all biological phenomena is, in principle, impossible.

1

Although the term "reduction" has come to be widely used in philosophical discussions of science, it has no standard definition. It is therefore not surprising that the term encompasses several sorts of things which need to be distinguished. But before I do this, a brief terminological excursion is desirable. Scientists and philosophers often talk of deducing or inferring one phenomenon from another (e.g., of deducing a planet's orbital motion), of explaining events or their concatenations (e.g., of explaining the occurrence of rainbows), and of reducing certain processes, things, or their properties to others (e.g., of reducing the process of heat conduction to molecular motions). However, these locutions are elliptical, and sometimes lead to misconceptions and confusions. For strictly speaking, it is not phenomena which are deduced from other phenomena, but rather *statements* about phenomena from other statements. This is obvious if we remind ourselves that a given phenomenon can be subsumed under a variety of distinct descriptions, and that phenomena make no assertions or claims. Consequently, until the traits or relations of a phenomenon which are to be discussed are indicated, and predications about them are formulated, it is literally impossible to make any deductions from them. The same holds true for the locutions of explaining or reducing phenomena. I will therefore avoid these elliptic modes of speech hereafter, and talk instead of deducing, explaining, or reducing statements about some subject matter.

Whatever else may be said about reductions in science, it is safe to say that they are commonly taken to be explanations, and I will so regard them. In consequence, I will assume that, like scientific explanations in general, every reduction can be construed as a series of statements, one of which is the conclusion (or the statement which is being reduced), while the others are the premises or reducing statements. Accordingly, reductions can be conveniently classified into two major types: homogeneous reductions, in which all of the "descriptive" or specific subject matter terms in the conclusion are either present in the premises also or can be explicitly defined using only terms that are present; and inhomogeneous reductions, in which at least one descriptive term in the conclusion neither occurs in the premises nor is definable by those that do occur in them. I will now characterize in a general way what I believe to be the main components and the logical structure of these two types of reduction, but will also state and comment upon some of the issues that have been raised by this account of reduction.

A frequently cited example of homogeneous reduction is the explanation by Newtonian mechanics and gravitational theory of various special laws concerning the motions of bodies, including Galileo's law for freely falling bodies near the earth's surface and the Keplerian laws of planetary motion. The explanation is homogeneous, because on the face of it at any rate, the terms occurring in these laws (e.g., distance, time, and acceleration) are also found in the Newtonian theory. Moreover, the explanation is commonly felt to be a reduction of those laws, in part because these laws deal with the motions of bodies in restricted regions of space which had traditionally been regarded as essentially dissimilar from motions elsewhere (e.g., terrestrial as contrasted with celestial motions), while Newtonian theory ignores this traditional classification of spatial regions and incorporates the laws into a unified system. In any event, the reduced statements in this and other standard examples of homogeneous reduction are commonly held to be deduced logically from the reducing premises. In consequence, if the examples can be taken as typical, the formal structure of homogenous reductions is in general that of deductive explanations. Accordingly, if reductions of this type are indeed deductions from theories whose range of application is far more comprehensive and diversified than that of the conclusions derived from them, homogeneous reductions appear to be entirely unproblematic, and to be simply dramatic illustrations of the well understood procedure of deriving theorems from assumed axioms.

However, the assumption that homogeneous reductions are deductive explanations has been recently challenged by a number of thinkers, on the ground that even in the stock illustrations of such reductions the reduced statements do not in general follow from the explanatory premises. For example, while Galileo's law asserts that the acceleration of a freely falling body near the earth's surface is constant, Newtonian theory entails that the acceleration is not constant, but varies with the distance of the falling body from the earth's center of mass. Accordingly, even though the Newtonian conclusion may be "experimentally indistinguishable" from Galileo's law, the latter is in fact "inconsistent" with Newtonian theory. Since it is this theory rather than Galileo's law that was accepted as sound, Galileo's law was therefore *replaced* by a different law for freely falling bodies, namely the law derived from the Newtonian assumptions. A similar outcome holds for Kepler's third planetary law. The general thesis has therefore been advanced that homogeneous reductions do not consist in the deduction or explanation of laws, but in the total *replacement* of incorrect assumptions by radically new ones which are believed to be more correct and precise than those they replace. This thesis raises far-reaching issues, and I will examine some of them presently. But for the moment I will confine my comments on it to questions bearing directly on homogeneous reductions.

a) It is undoubtedly the case that the laws derivable from Newtonian theory do not coincide exactly with some of the previously entertained hypotheses about the motions of bodies, though in other cases there may be such coincidence. This is to be expected. For it is a widely recognized function of comprehensive theories (such as the Newtonian one) to specify the conditions under which antecedently established regularities hold, and to indicate, in the light of those conditions, the modifications that may have to be made in the initial hypotheses, especially if the range of application of the hypotheses is enlarged. Nevertheless, the initial hypotheses may be reasonably close approximations to the consequences entailed by the comprehensive theory, as is indeed the case with Galileo's law as well as with Kepler's third Law. (Incidentally, when Newtonian theory is applied to the motions of just two bodies, the first and second Keplerian laws agree fully with the Newtonian conclusions). But if this is so, it is correct to say that in homogeneous reductions the reduced laws are either derivable from the explanatory premises, or are good approximations to the laws derivable from the latter.

b) Moreover, it is pertinent to note that in actual scientific practice, the derivation of laws from theories usually involves simplifications and approximations of various kinds, so that even the laws which are allegedly entailed by a theory are in general only approximations to what is strictly entailed by it. For example, in deriving the law for the period of a simple pendulum, the following approximative assumptions are made: the weight of the pendulum is taken to be concentrated in the suspended bob; the gravitational force acting on the bob is assumed to be constant, despite variations in the distance of the bob from the earth's center during the pendulum's oscillation; and since the angle through which the pendulum may oscillate is stipulated to be small, the magnitude of the angle is equated to the sine of the angle. The familiar law that the period of a pendulum is proportional to the square root of its length divided by the constant of acceleration is therefore derivable from Newtonian theory only if these various approximations are taken for granted.

More generally, though no statistical data are available to support the claim, there are relatively few deductions from the mathematically formulated theories of modern physics in which analogous approximations are not made, so that many if not all the laws commonly said by scientists to be deducible from some theory are not strictly entailed by it. It would nevertheless be an exaggeration to assert that in consequence scientists are fundamentally mistaken in claiming to have made such deductions. It is obviously important to note the assumptions, including those concerning approximations, under which the deduction of a law is made. But it does not follow that given those assumptions a purported law cannot count as a consequence of some theory. Nor does it follow that if in a proposed homogeneous reduction the allegedly reduced law is only an approximation to what is actually entailed by the reducing theory when *no* approximative assumptions are added to the latter, the law has not been reduced but is being replaced by a radically different one.

c) Something must also be said about those cases of homogeneous reduction in which the law actually derivable from the reducing theory makes use of concepts not employed in the law to be reduced. Thus, while according to Kepler's third (or harmonic) law, the squares of the periods of the planets are to each other as the cubes of their mean distances from the sun, the Newtonian conclusion is that this ratio is not constant for all the planets but varies with their *masses*. But the notion of mass was introduced into mechanics by Newton, and does not appear in the Keplerian law; and although the masses of the planets are small in comparison with the mass of the sun, and the Keplerian harmonic law is therefore a close approximation to the Newtonian one, the two cannot be equated. Nevertheless, while the two are not equivalent, neither are they radically disparate in content or meaning. On the contrary, the Newtonian law identifies a causal factor in the motions of the planets which was unknown to Kepler.

2

I must now turn to the second major type of reductive explanations. Inhomogeneous reductions, perhaps more frequently than homogenous ones, have occasioned vigorous controversy among scientists as well as philosophers concerning the cognitive status, interpretation, and function of scientific theories; the relations between the various theoretical entities postulated by these theories, and the familiar things of common experience; and the valid scope of different modes of scientific analysis. These issues are interconnected, and impinge in one way or another upon questions about the general structure of inhomogenous reductions. Since none of the proposed answers to these issues has gained universal assent, the nature of such reductions is still under continuing debate.

Although there are many examples of inhomogeneous reductions in the history of science, they vary in the degree of completeness with which the reduction has been effected. In some instances, all the assumed laws in one branch of inquiry are apparently explained in terms of a theory initially developed for a different class of phenomena; in others, the reduction has been only partial, though the hope of completely reducing the totality of laws in a given area of inquiry to some allegedly "basic" theory may continue to inspire research. Among the most frequently cited illustrations of such relatively complete inhomogenous reductions are the explanation of thermal laws by the kinetic theory of matter, the reduction of physical optics to electromagnetic theory, and the explanation (at least in principle) of chemical laws in terms of quantum theory. On the other hand, while some processes occurring in living organisms can now be understood in terms of physicochemical theory, the reducibility of all biological laws in a similar manner is still a much disputed question.

In any case, the logical structure of inhomogeneous reductive explanations is far less clear and is more difficult to analyze than is the case with homogeneous reductions. The difficulty stems largely from the circumstance that in the former there are (by definition) terms or concepts in the reduced laws (e.g., the notion of heat in thermodynamics, the term "light-wave" in optics, or the concept of valence in chemistry) which are absent from the reducing theories. Accordingly, if the overall structure of the explanation of laws is taken to be that of a deductive argument, it seems impossible to construe inhomogeneous reductions as involving essentially little more than the logical derivation of the reduced laws (even when qualifications about the approximative character of the latter are made) from their explanatory premises. If inhomogeneous reductions are to be subsumed under the general pattern of scientific explanations, it is clear that additional assumptions must be introduced as to how the concepts characteristically employed in the reduced laws, but not present in the reducing theory, are connected with the concepts that do occur in the latter.

Three broad types of proposals for the structure of inhomogeneous reductions can be found in the recent literature of the philosophy of science. The first, which for convenience will be called the "instrumentalist" analysis, is usually advocated by thinkers who deny a cognitive status to scientific laws or theories, regarding them as neither true nor false but as rules (or "inference tickets") for inferring so-called "observation statements" (statements about particular events or occurrences capable of being "observed" in some not precisely defined sense) from other such statements. According to this view, for example, the kinetic theory of gases is not construed as an account of the composition of gases. It is taken to be a complex set of rules for predicting, among other things, what the pressure of a given volume of gas will be if its temperature is kept constant but its volume is diminished. However, the scope of application of a given law or theory may be markedly more limited than the scope of another. The claim that a theory T (e.g., the corpus of rules known as thermodynamics) is reduced to another theory T′ (e.g., the kinetic theory of gases) would therefore be interpreted as saying that all the observation statements which can be derived from given data with the help of T can also be derived with the help of T′, but not conversely. Accordingly, the question to which this account of inhomogeneous reduction is addressed is not the ostensibly asserted content of the theories involved in reduction, but the comparative ranges of observable phenomena to which two theories are applicable.

Although this proposed analysis calls attention to an important function of theories and provides a rationale for the reduction of theories, its adequacy depends on the plausibility of uniformly interpreting general statements in science as rules of inference. Many scientists certainly do not subscribe to such an interpretation, for they frequently talk of laws as true and as providing at least an approximately correct account of various relations of dependence among things. In particular, this interpretation precludes the explanation of macro-states of objects in terms of unobservable micro-processes postulated by a theory. Moreover, the proposal is incomplete in a number of ways: it has nothing to say about how theoretical terms in laws (e.g., "electron" or even "atom") may be used in connection with matters of observation, or just how theories employing such notions operate as rules of inference; and it ignores the question of how, if at all, the concepts of a reduced theory are related to those of the reducing one, or in what way statements about a variety of observable things may fall within the scope of both theories. In consequence, even if the proposed analysis is adequate for a limited class of reductive explanations, it does not do justice to important features characterizing many others.

The second proposed analysis of inhomogeneous reductions (hereafter to be referred to—perhaps misleadingly—as the "correspondence" proposal) is also based on several assumptions. One of them is that the terms occurring in the conclusion but not in the premises of a reduction have "meanings" (i.e., uses and applications) which are determined by the procedures and definitions of the discipline to which reduced laws initially belong, and can be understood without reference to the ideas involved in the theories to which the laws have been reduced. For example, the term "entropy" as used in thermodynamics is defined independently of the notions characterizing statistical mechanics. Furthermore, the assumption is made that many subject matter terms common to both the reduced and reducing theories—in particular, the so-called observation terms employed by both of them to record the outcome of observation and experiment—are defined by procedures which can be specified independently of these theories and, in consequence, have "meanings" that are neutral with respect to the differences between the theories. For example, the terms "pressure" and "volume change" which occur in both thermodynamics and the kinetic theory of gases are used in the two theories in essentially the same sense. It is important to note, however, that this assumption is compatible with the view that even observation terms are "theory impregnated," so that such terms are not simply labels for "bare sense-data," but predicate characteristics that are not immediately manifest and are defined on the basis of various theoretical commitments. For example, if the expression "having a diameter of five inches" is counted as an observation predicate, its application to a given object implicitly involves commitment to some theory of spatial measurement as well as to some laws concerning the instrument used in making the measurement. Accordingly, the point of the assumption is not that there are subject-matter terms whose meanings or uses are independent of *all* theories, but rather that every such term has a meaning which is fixed by *some* theory but independent of others. A third assumption underlying the correspondence analysis of inhomogeneous reductions is that, like homogeneous reduction, and with similar qualifications referring to approximations, they embody the pattern of deductive explanations.

In view of these assumptions, it is clear that if a law (or theory) T is to be reduced to a theory T′ not containing terms occurring in T, T′ must be supplemented by what have been called "rules of correspondence" or "bridge laws," which establish *connections* between the distinctive terms of T and certain terms (or combinations of terms) in T′. For example, since the second law of thermodynamics talks of the transfer of heat, this law cannot be deduced from classical mechanics, which does not contain the term "heat," unless the term is connected in some way with some complex of terms in mechanics. The statement of such a connection is a correspondence rule. However, because of the first of the above three asummptions, a correspondence rule cannot be construed as an explicit definition of a term distinctive of T, which would permit the elimination of the term on *purely logical grounds* in favor of the terms in T′. Thus, the notion of entropy as defined in thermodynamics can be understood and used without any reference to notions employed in theories about the microstructure of matter; and no amount of logical analysis of the concept of entropy can show the concept to be constituted out of the ideas employed in, say, statistical mechanics. If this is indeed the case (as I believe it is), then the theory T is not derivable from (and hence not reducible to) the theory T′, although T may be derivable from T′ when the latter is conjoined with an appropriate set of bridge laws.

What then is the status of the correspondence rules required for inhomogeneous reduction? Different articulations of the theories involved in a reduction, as well as different stages in the development of inquiry into the subject matter of the theories, may require different answers; but I will ignore these complications. In general, however, correspondence rules formulate *empirical hypotheses*—hypotheses which state certain relations of dependence between things mentioned in the reduced and reducing theories. The hypotheses are, for the most part, not testable by confronting them with observed instances of the relations they postulate. They are nevertheless not arbitrary stipulations, and as with many other scientific laws their factual validity must be assessed by comparing various consequences entailed by the system of hypotheses to which they belong with the outcome of controlled observations. However, bridge laws have various forms; and while no exhaustive classification of their structure is available, two sorts of bridge laws must be briefly described.

a) A term in a reduced law may be a predicate which refers to some distinctive *attribute* or characteristic of things (such as the property of having a certain temperature or of being red) that is not denoted by any of the predicates of the reducing theory. In this case the bridge law may specify the conditions, formulated in terms of the ideas and assumptions of the reducing theory, under which the attribute occurs. For example, the kinetic theory of gases formulates its laws in terms of such notions as molecule, mass, and velocity, but does not employ the thermodynamical notion of temperature. However, a familiar bridge law states that a gas has a certain temperature when the mean kinetic energy of its molecules has a certain magnitude. In some cases, bridge laws of the sort being considered may specify conditions for the occurrence of an attribute which are necessary as well as sufficient; in other cases the conditions specified may be sufficient without being necessary; and in still other cases, the conditions stated may only be necessary. In the latter case, however, laws involving the attribute will, in general, not be deducible from the proposed reducing theory. (Thus, though some of the necessary conditions for objects having colors can be stated in terms of ideas belonging to physical optics in its current form, the physiological equipment of organisms which must also be present for the occurrence of colors cannot be described in terms of those ideas. Accordingly, if there are any laws about color relations, they are not reducible to physical optics.)

In any case, such bridge laws are empirical hypotheses concerning the *extensions* of the predicates mentioned in these correspondence rules—that is, concerning the classes of individual things or processes designated by those predicates. An attribute of things connoted by a predicate in a reduced law may indeed be quite different from the attribute connoted by the predicates of the reducing theory; but the class of things possessing the former attribute may nevertheless coincide with (or be included in) the class of things which possess the property specified by a complex predicate in the reducing theory. For example, the statement that a liquid is viscous is not equivalent in meaning to the statement that there are certain frictional forces between the layers of molecules making up the liquid. But if the bridge laws connecting the macro-properties and the microstructure of liquids is correct, the extension of the predicate "viscous" coincides with (or is included in) the class of individual systems with that microstructure.

b) Let me now say something about a second sort of correspondence rule. Although much scientific inquiry is directed toward discovering the determining conditions under which various traits of things occur, some of its important achievements consist in showing that things and processes initially assumed to be distinct are in fact the same. A familiar example of such an achievement is the discovery that the Morning Star and the Evening Star are not different celestial objects but are identical. Similarly, although the term "molecule" designates one class of particles and the term "atom" designates another class, molecules are structures of atoms, and in particular a water molecule is an organization of hydrogen and oxygen atoms denoted by the formula "H_2O"; and accordingly, the extension of the predicate "water molecule" is the same as the class of things designated by the formula. Correspondence rules of the second sort establish analogous identifications between classes of individuals or "entities" (such as spatiotemporal objects, processes, and forces) designated by different predicates. An oft cited example of such rules is a bridge law involved in the reduction of physical optics to electromagnetic theory. Thus, prior to Maxwell, physicists postulated the existence of certain physical propagations designated as "light waves," while electromagnetic theory was developed on the assumption that there are electromagnetic waves. An essential step in the reduction of optics to electrodynamics was the introduction by Maxwell of the hypothesis (or bridge law) that these are not two *different* processes but a *single* one, even though electromagnetic waves are not always manifested as visible light. Analogous bridge laws are assumed when a flash of lightning is said to be a surge of electrically charged particles, or when the evaporation of a liquid is asserted to be the escape of molecules from its surface; and while the full details for formulating a similar bridge law are not yet available, the hope of discovering them underlies the claim that a biological cell is a complex organization of physicochemical particles.

Correspondence rules of the second kind thus differ from rules of the first, in that unlike the latter (which state conditions, often in terms of the ideas of a micro-theory, for the occurrence of traits characterizing various things, often macroscopic ones), they assert that certain logically nonequivalent expressions describe identical entities. Although both sorts of rules have a common function in reduction and both are in general empirical assumptions, failure to distinguish between them is perhaps one reason for the persistence of the mistaken belief that reductive explanations establish the "unreality" of those distinctive traits of things mentioned in reduced laws.

3

This account of inhomogeneous reduction has been challenged by a number of recent writers who have advanced an alternate theory which rejects the main assumptions of both the instrumentalist and the correspondence analyses, and which I will call the "replacement" view. Since I believe the correspondence account to be essentially correct, I shall examine the fundamental contention of the replacement thesis, as presented by Professor Paul Feyerabend, one of its most vigorous proponents.

Feyerabend's views on reduction rest upon the central (and on the face of it, sound) assumption that "the meaning of every term we use depends upon the theoretical context in which it occurs."[1] This claim is made not only for "theoretical" terms like "neutrino" or "entropy" in explicitly formulated scientific theories, but also for expressions like "red" or "table" used to describe matters of common observation (i.e., for observation terms). Indeed, Feyerabend uses the word "theory" in a broad sense, to include such things as myths and political ideas.[2] He says explicitly that "even everyday languages, like languages of highly theoretical systems, have been introduced in order to give expression to some theory or point of view, and they therefore contain a well-developed and sometimes very abstract ontology."[3] "The description of every single fact," he declares, is "dependent on *some* theory."[4] He further maintains that "theories are meaningful independent of observations; observational statements are not meaningful unless they have been connected with theories."[5] There is, therefore, no "observation core," even in statements of perception, that is independent of theoretical interpretation,[6] so that strictly speaking each theory determines its own distinctive set of observation statements. And while he allows that two "low level" theories which fall within the conceptual framework of a comprehensive "background theory" may have a common interpretation for their observation statements, two "high level" theories concerning the nature of the basic elements of the universe "may not share a single observational statement."[7] It is therefore allegedly an error to suppose that the empirical adequacy of a theory can be tested by appeal to observation statements whose meanings are independent of the theory, and which are neutral as between that theory and some alternative competing theory. "The methodological unit to which we must refer when discussing questions of test and empirical context, is constituted by a *whole set of partly overlapping, factually adequate, but mutually inconsistent theories*."[8]

Moreover, a change in a theory is accompanied by a change in the meanings of all its terms, so that theories constructed on "mutually inconsistent principles" are in fact "incommensurable."[9] Thus, if T is classical celestial mechanics, and T′ is the general theory of relativity, "the meanings of all descriptive terms of the two theories, primitive as well as defined terms, will be different," the theories are incommensurable, and "not a single descriptive term of T can be incorporated into T′."[10] In consequence, Feyerabend believes the correspondence account of inhomogeneous reduction is basically mistaken in supposing that allegedly reduced laws or theories can be derived from the reducing theory with the help of appropriate bridge laws:

> What happens . . . when transition is made from a theory T′ to a wider theory T (which, we shall assume, is capable of covering all the phenomena that have been covered by T′) is something much more radical than incorporation of the *unchanged* theory T′ (unchanged, that is, with respect to the meanings of its main descriptive terms as well as to the meanings of the terms of its observation language) into the context of T. What does happen is, rather, a *complete replacement* of the ontology (and perhaps even of the formalism) of T′ by the ontology (and the formalism) of T and a corresponding change of the meanings of the descriptive elements of the formalism of T′ (provided these elements and this formalism are still used). This replacement affects not only the theoretical terms of T′ but also at least some of the observational terms which occurred in its test statements. . . . In short: introducing a new theory involves changes of outlook both with respect to the observable and with respect to the unobservable features of the world, and corresponding changes in the meaning of even the most "fundamental" terms of the language employed.[11]

Accordingly, if these various claims are warranted, there is not and cannot be any such thing as the reduction of laws or theories; and the examples often cited as instances of reduction are in fact instances of something else: the exclusion of previously accepted hypotheses from the corpus of alleged scientific knowledge, and the substitution for them of incommensurably different ones.

But are these claims warranted? I do not believe they are. Feyerabend is patently sound in maintaining that no single statement or any of its constituent terms has a meaning in isolation, or independently of various rules or conventions governing its use. He is no less sound in noting that the meaning of a word may change when its range of application is altered. However, these familiar truisms do not support the major conclusion he draws from them. The presentation of his thesis suffers from a number of unclarities (such as what is to count as a change in a theory, or what are the criteria for changes in meaning), which cloud the precise import of some of his assertions. I shall, however, ignore these unclarities here[12] and will comment briefly only on two difficulties in Feyerabend's argument.

a) It is a major task of scientific inquiry to assess the adequacy of proposed laws to the "facts" of a subject matter as established by observation or experiment, and to ascertain whether the conclusions reached are consistent with one another. However, if two proposed theories for some given range of phenomena share no term with the same meaning in each of them, so that the theories have completely different meanings (as Feyerabend believes is commonly the case), it is not evident in what sense two such theories can be said to be either compatible or inconsistent with one another: for relations of logical opposition obtain only between statements whose terms have common meanings. Moreover, it is also difficult to understand how, if the content of observation statements is determined by the theory which is being tested (as Feyerabend maintains), those statements can serve as a basis for deciding between the theory and some alternative to it. For according to his analysis those observation statements will automatically corroborate the theory that happens to be used to interpret observational data, but will be simply irrelevant in assessing the empirical validity of an alternative theory. Theories thus appear to be self-certifying, and to be beyond the reach of criticism based on considerations that do not presuppose them. This outcome is reminiscent of Karl Mannheim's claim that truth in social matters is "historically relative": there are no universally valid analyses of social phenomena, since every such analysis is made within some distinctive social perspective which determines the meaning as well as the validity of what is said to be observed, so that those who do not share the same perspective can neither reach common conclusions about human affairs, nor significantly criticize each others' findings.

Feyerabend attempts to escape from such skeptical relativism by involving what he calls the "pragmatic theory of observation." In this theory, it is still the case that the meaning of an observation statement varies with the theory used to interpret observations. However, it is possible to describe the observational and predictive statements an investigator utters as *responses* to the situations which "prompt" the utterances, and to compare the order of these responses with the order of the physical situations that prompt them, so as to ascertain the agreements or disagreements between the two orders.[13] But if this account of the role of observation statements in testing theories is to outflank the relativism Feyerabend wants to avoid, the *secondary* statements (they are clearly observation statements) about the responses (or primary observation statements) of investigators cannot have meanings dependent on the theory being tested, and must be invariant to alternative theories. However, if secondary statements have this sort of neutrality, it is not evident why only such observation statements can have this privileged status.

b) Feyerabend has difficulties in providing a firm observational basis for objectively evaluating the empirical worth of proposed hypotheses. The difficulties stem from what I believe is his exaggerated view that the meaning of every term occurring in a theory or in its observation statements is wholly and uniquely determined by that theory, so that its meaning is radically changed when the theory is modified. For theories are not quite the monolithic structures he takes them to be—their component assumptions are, in general, logically independent of one another, and their terms have varying degrees of dependence on the theories into which they enter. Some terms may indeed be so deeply embedded in the totality of assumptions constituting a particular theory that they can be understood only within the framework of the theory: e.g., the meaning of "electron spin" appears to be inextricably intertwined with the characteristic ideas of quantum theory. On the other hand, there are also terms whose meanings seem to be invariant in a number of different theories: e.g., the term "electric charge" is used in currently accepted theories of atomic structure in the same sense as in the earlier theories of Rutherford and Bohr. Similar comments apply to observation terms, however these may be specified. Accordingly, although both "theoretical" and "observational" terms may be "theory laden," it does not follow that there can be no term in a theory which retains its meaning when it is transplanted into some other theory.

More generally, it is not clear how, on the replacement view of reduction, a theory T can be at the same time more inclusive than, and also have a meaning totally different from, the theory T′ it allegedly replaces—especially since according to Feyerabend the replacing theory will entail "that all the concepts of the preceding theory have extension zero, or . . . it introduces rules which cannot be interpreted as attributing specific properties to objects within already existing classes, but which change the system of classes itself."[14] Admittedly, some of the laws and concepts of the "wider theory" often differ from their opposite numbers in the earlier theory. But even in this case, the contrasted items may not be "incommensurable." Thus, the periodic table classifies chemical elements on the basis of certain patterns of similarity between the properties of the elements. The description (or theoretical explanation) of those properties has undergone important changes since the periodic table was first introduced by Mendeleev. Nevertheless, though the descriptions differ, the classification of the elements has remained fairly stable, so that fluorine, chlorine, bromine, and iodine, for example, continue to be included in the same class. The new theories used in formulating the classification certainly do not entail that the concepts of the preceding ones have zero extension. But it would be difficult to understand why this is so if, because of differences between the descriptions, the descriptions were totally disparate.

Consider, for example, the argument that thermodynamics is not reducible to statistical mechanics, on the ground that (among other reasons) entropy is a statistical notion in the latter theory but not in the former one: since the meaning of the word "entropy" differs in the two theories, entropy laws in statistical mechanics are not derivable from entropy laws in thermodynamics (and in fact are said to be incompatible). Admittedly, the connotation of the word "entropy" in each of the two theories is not identical; and if the correspondence account of reduction were to claim that they are the same, it would be patently mistaken. But the fact remains that the two theories deal with many phenomena common to both their ranges; and the question is how is this possible? In brief, the answer seems to be as follows. The word "entropy" in thermodynamics is so defined that its legitimate application is limited to physical systems satisfying certain specified conditions, e.g., to systems such as gases, whose internal motions are not too "tumultuous" (the word is Planck's), a condition which is not satisfied in the case of Brownian motions. These conditions are relaxed in the definition of "entropy" in statistical mechanics, so that the extension of the Boltzmann notion of entropy includes the extension of the Clausius notion. In consequence, despite differences in the connotations of the two definitions, the theories within which they are formulated have a domain of application in common, even though the class of systems for which thermodynamical laws are approximately valid is more restricted than is the class for the laws of statistical mechanics. But it is surely not the case that the latter theory implies that the Clausius definition of entropy has a zero extension or that the laws of thermodynamics are valid for no physical systems whatsoever.

This difficulty of the replacement view in explaining how the "wider" theory, which allegedly replaces a "narrower" one, may nevertheless have a domain of common application, does not arise in the correspondence account of reduction. For the bridge laws upon which the latter sets great store are empirical hypotheses, not logically true statements in virtue of the connotations of the terms contained in them. Bridge laws state what relations presumably obtain between the *extensions* of their terms, so that in favorable cases laws of the "narrower" theory (with suitable qualifications about their approximate character) can be deduced from the "wider" theory, and thereby make intelligible why the two theories may have a common field of application. Accordingly, although I will not pretend that the correspondence account of reduction is free from difficulties or that I have resolved all of them, on the whole it is a more adequate analysis than any available alternative to it.

4

Let me now turn to the current controversy over the reducibility of biological laws to physical ones, in the hope that the above considerations may throw some light on the issues of the debate. Despite the remarkable advances in the preceding decades in discovering physicochemical mechanisms involved in living processes, no one disputes the fact that at present, physicochemical explanations for *all* biological laws are not available. However, while some outstanding biologists believe that such explanations will eventually be forthcoming, others (who also reject vitalistic doctrines) deny that the reduction of all biological laws to physical ones is possible. Various reasons for this denial have been given by a number of recent writers (e.g., by Professors Barry Commoner, Walter Elsasser, Bentley Glass, and Michael Polanyi, among others), and all of these reasons deserve careful attention. However, I must restrict myself here to examining only the important double-pronged argument presented by Bentley Glass.[15]

a) Glass begins the first part of his argument by observing that "random" behavior (behavior which eventuates in an ordered distribution of properties exhibiting statistical rather than uniform regularities that are formulated by statistical laws) is found at all levels of organized matter. For example, the Mendelian laws express statistical regularities resulting from the equal probability of an egg's fertilization by each one of several kinds of sperm; and the Hardy-Weinberg law states the statistical regularities arising from the random mating of individuals in a population containing different genotypes. However, while both laws express regularities in the transmission of genes from one generation to another, the former is at the cell level of organization while the latter is at the level of interbreeding populations. But Glass also believes that the randomness of the units at one level of organization "does not necessarily depend" on the randomness of the units at lower levels.[16] On this important assumption, he therefore maintains that neither law is derivable from the other.[17] More generally, he concludes "[S]tatistical laws of one level of organization are not reducible to the statistical laws of another."[18] In consequence, although physical laws which are established for nonliving systems also hold in living ones, they cannot explain all the laws of the latter.

It is clear, however, that the cogency of the argument depends on the validity for a *given set* of levels of organization of what I called Glass's "important assumption." But let me first note that the argument would have the same force if, instead of supposing that the laws at different levels are statistical, the laws were supposed to be strictly universal or "deterministic"; e.g., if it were supposed that eggs are fertilized by sperm in accordance with a deterministic law L, and individuals mated in accordance with a deterministic law L'. For even on this hypothesis, L' would not be derived from L, *unless* some bridge law were available which stated sufficient conditions for the mating of individuals in terms of the fertilization of eggs. However, it is not *logically impossible* that such connections exist, and the question whether they do or not falls into the province of empirical inquiry rather than of *a priori* reasoning.

Similarly, Glass is certainly correct in declaring that randomness at one level of organization does not necessarily depend on randomness at a lower level, so that his "important assumption" denies, in effect, the *availability* of certain bridge laws needed for reduction, not their *possibility*. Accordingly, his argument for the underivability of biological from physical laws is based on the present state of scientific knowledge, and does not show that physical laws cannot "be expected ultimately to explain *all* the laws of living systems."[19] However unlikely the establishment of the required bridge laws may seem at present, the argument does not altogether rule out the reduction of biological to physical laws. After all, the macroscopic (or "higher level") regularities found in chemical interactions are reducible to the microscopic statistical regularities formulated by quantum theory.

b) The second argument Glass offers for doubting the reducibility of biological to physical laws rests on the "uniqueness" of living organisms and the "indeterminacy" of the evolutionary process. One reason he mentions for the uniqueness of organisms is that in a sexually reproducing but not strictly inbred population, the genotype of each individual is not likely to recur throughout history. Accordingly, even statistical prediction is possible only if certain common characteristics are abstracted from "the infinitely varied individuals,"[20] and only if the populations and samples are sufficiently large (a condition infrequently realized in biological study). On the other hand, the major reason he gives for the indeterminacy of the evolutionary process is that evolution takes place because of genetic mutations, whose possible kinds may recur with certain relative frequencies, but whose actual occurrences are unpredictable. In consequence, the evolutionary history of a population is determined by the generally unpredictable occurrences of random events, so that biological explanations "can in some respects not be reduced to the laws of physical science."[21]

This argument, like the first one Glass presents, calls attention to features of biological laws that are often neglected in discussions of their reducibility to physics. Nevertheless, I do not think it is quite as "unanswerable" as he believes it to be:

i) In the first place, there appears to be a conflation of two questions that need to be distinguished: a) whether evolutionary developments or other biological phenomena can be predicted, and b) whether biological laws can be *reduced* to physical ones. An event is commonly said to be predictable (whether or not with maximum probability) if, and only if, two conditions are satisfied: a law (or set of laws) must be available which states the conditions for the occurrence of the event; and the initial and boundary conditions for the application of the law to a given instance must be known. Accordingly, one reason for our inability to predict an event may be that the requisite initial and boundary conditions are not fully known prior to the occurrence of the event, even though they may be ascertained subsequent to its occurrence and the event explained in retrospect. In this case, the possibility of reducing the law to another is obviously not excluded. A quite different reason for such inability may be that there are no known laws either for the event's occurrence, or for the occurrence of those of its properties in which we are especially interested. In this case, however, the unpredictability of the event is not relevant to the issue of reduction, since there is no candidate for reduction to some other laws. In either case, therefore, the indeterminacy of the evolutionary process and the limited value of evolutionary theory as an instrument for prediction do not count against the possibility of reducing the theory to physical laws.

ii) In the second place, though living organisms may be unique, so may many actually existing physical systems, especially if they are constituted out of numerous components. No two stars, or two specimens of quartz, or even two watches manufactured by some standardized process, are precisely alike in their properties or behaviors; and while we may often regard them as "essentially" alike, we are in fact ignoring differences of which we may be unaware, or which for one reason or another we think are unimportant. But in any case, science, like discursive thought in general, cannot deal with things insofar as they are unique; and as Dr. Glass makes clear, laws can be formulated for biological organisms only by prescinding from the combinations of characteristics unique to each of them certain traits they have in common. In this respect, however, all branches of science are in the same boat, and nothing seems to follow from the special uniqueness of living organisms that bears on the reducibility of biological laws.

But Glass also notes that in biology, in contrast to what is generally the case in the physical sciences, the populations of individuals selected for study—that is, the unique organisms grouped together because they possess various characteristics in common which are believed to have theoretical significance—are frequently far too small to exhibit statistical regularities, or to permit the use of statistical laws for highly accurate predictions. This is indeed a difficulty. But it is an obstacle to the *establishment* of biological laws and the *forecasting* of evolutionary developments; it is not an objection to the possibility of *reducing* whatever biological laws may be available to physical ones.

Accordingly, I do not think that Glass has demonstrated the irreducibility of biology to physics—indeed, I do not believe that such a demonstration can be given, in part because physics is still a developing science. He has nevertheless given ample reasons for suspended judgment on the claim now fashionable with many microbiologists that the laws of physics *as currently constituted* comprehend all the laws of biology. For he has pointed out that bridge laws to connect biological with physical characteriestics at different levels of biological organization are still lacking and are indispensable for the reduction of biology. He has also noted some of the serious difficulties that must be overcome if such bridge laws are to be established. Moreover, he has made clear—and I regard this as especially salutary at a time such as ours when reductive tendencies threaten to obliterate important distinctions—that although physicochemical analyses of biological processes have made, and will doubtless continue to make, enormous contributions to our understanding of biological phenomena, the reduction of biology to physics is not a necessary condition for the advancement of biological knowledge.

7/Carnap's Theory of Induction

C. D. Broad once remarked that though inductive reasoning is the glory of science, it is the scandal of philosophy. Whether or not this characterization of philosophy is a merited one, there is no doubt that despite substantial advances made by logicians and philosophical scientists in the analysis of inductive arguments, even competent students continue to disagree on many fundamental issues encountered in the subject. These issues include not only the notorious general problem of "justifying" principles of inductive reasoning, but also special questions concerning the formal logic and the methodology of inductive inference. They run the gamut from doubts about the relevance of the mathematical calculus of probability to the task of codifying the tacit rules governing habitual inductive reasoning, through questions about the conditions under which inductive arguments are valid and about the correct analysis of the central notion of "the weight of evidence," to problems concerning the epistemic status of generally accepted principles of inductive inference. If it is a scandal to have unresolved issues, then the present state of philosophic discussion on induction is indeed scandalous.

What is perhaps Carnap's most ambitious contribution to logical analysis is his monumental but incompleted attempt to put an end to much of this scandal, if scandal it is. He has set himself the important task of codifying the logic of induction, in a manner analogous to modern systematizations of deductive logic, and of doing this within the unitary framework provided by a precise quantitative explication of the basic idea of "the strength of evidential support." The foundations for his system have been laid deep and in a characteristically meticulous fashion; and though the structure is far from complete, its present outlines already exhibit the magnificent architectonic qualities of the completed design. Carnap employs his basic conceptions with brilliant ingenuity, and he discusses mooted questions in the philosophy of induction with flexible insight and with his usual candor. No students of the problems of inductive inference, whether or not they find themselves in agreement with Carnap's approach, can fail to be instructed by the comprehensive analyses that support his system.

In this essay I propose to evaluate Carnap's theory of induction, even though such an evaluation is perhaps premature. It may be premature, because the published form of the theory is still but a fragment, and develops in detail only a relatively small portion of its intended content. Indeed, except for some preliminary outlines, a full explication of the notion of evidential support which Carnap apparently favors is not yet available; and it is not even entirely clear which, if any, of the infinitely many inductive methods that he presents as possible ones he will eventually recommend as the most promising. Comments on the theory at the present stage of its development may therefore turn out to be quite pointless in the light of the subsequent elaborations the theory will doubtless receive. Moreover, Carnap has reserved for later discussion a number of basic questions affecting the applicability of his theory. He has in fact been primarily engaged thus far in constructing a *logic* of inductive inference—in developing deductively relations between statements, each of which ascribes a degree of "confirmation" (or evidential support) to a given hypothesis by given evidence, and each of which is logically certifiable in the light of the postulates and definitions adopted. He has thus far not given comparable attention (at least not in published form) to what he calls the "methodological" problems that are generated when the *applicability* of the logic to actual scientific procedure is considered. Carnap's *logic* of induction, like any branch of pure mathematics, can of course be developed and examined without reference to its possible uses in empirical inquiry; but its worth as a *theory* of induction—as an explication and refined extension of ideas and principles employed in the search for empirical truth—cannot be judged independently of such reference. As Carnap makes clear, his definition of "degree of confirmation" and his theorems in inductive logic are intended to be reconstructions of familiar though vague notions and types of arguments implicit in the practice of science; and his logic is essentially a *proposal* that evidence for hypotheses be weighed in accordance with the rules which the logic postulates. But such a proposal cannot be evaluated exclusively in terms of the internal coherence of the system. Carnap himself notes that a deductive system "may be a theory which is wonderful to look at in its exactness, symmetry, and formal elegance, and yet woefully inadequate for the task of application for which it is intended."[1] Partly because of the unfinished state of Carnap's system, however, there have been no serious attempts to employ it in the conduct of inquiry; and objective data are therefore largely lacking for judging the adequacy of his inductive logic for its ostensible purpose.

Despite these obstacles, there is nevertheless some basis for venturing a critique of Canap's theory. He has developed enough of it to make clear the main lines of his approach; and though there is a scarcity of fully reliable evidence as to its potential effectiveness, there is much competent information concerning inductive practice, even if the use of such information for evaluating the theory must be somewhat impressionistic. It is perhaps unnecessary to add, however, that in view of the obstacles mentioned, a critique of Carnap's ideas on induction can at present be only a tentative one.

1

The stimulus for constructing an inductive logic is supplied by familiar facts such as the following: statements are frequently accepted in empirical inquiry as well-founded, even though the evidence for a given conclusion does not formally imply the latter; again, though the evidence for a hypothesis may not be regarded as sufficient to warrant its acceptance, the hypothesis may be judged to receive better (or stronger) support from one set of evidential premises than from another set; moreover, one hypothesis is sometimes taken to be better supported by given evidence than is some other hypothesis by the same or by different evidence. In such inductive arguments, the hypothesis (or conclusion) may be either general (strictly universal, existential, or statistical) or singular in form; they may employ only such notions that refer to matters accessible to direct observation, or they may use "theoretical" notions not applicable to directly observable things; and the evidential statements (or premises) may differ among themselves in a similar fashion. But a common feature of inductive arguments, setting them off from deductive ones, is that they may be "valid" or possess various degrees of "probative force," even though their conclusions are discovered to be false while their premises are assumed to be true.

It continues to be a matter of dispute whether in current practice numerical measures are *ever* assigned to the degree of support that available evidence lends to a hypothesis. For in those cases where such measures are apparently assigned (as in bets placed on the outcome of games of chance), students are in disagreement on the question whether the numerical values adopted are to be construed as measures of the weight of the evidence for a given hypothesis, or whether on the contrary those numerical values are to be understood as measures of the relative frequency with which *similar* hypotheses are true in classes of *similar* cases. On the other hand, no one seriously disputes the fact that in *most* cases current procedures in the sciences as well as in the practical affairs of life do not estimate the strength of evidential support in quantitative terms. It would be generally conceded, moreover, that there are many situations in which *no* degree of weight whatever is commonly associated with purported evidence for a given hypothesis, or in which the evidential support for one hypothesis is judged to be simply incomparable with the evidential support for another hypothesis. For example, if we assume that the captain of a certain ship was born in 1900, the hypothesis that he was 40 years old in 1959 is disconfirmed by this evidence; and if we were to contemplate introducing numerical measures, we might conceivably assign a zero degree of support to the evidence for that hypothesis. On the other hand, since information about the ship's position at sea on a certain day is irrelevant to the hypothesis about the captain's age, I do not believe *any* degree of support whatever (and certainly neither zero nor even one-half) would be assigned, in conformity with established habits of estimating evidence, to that "evidence" for the stated hypothesis. And *a fortiori,* we would not assign any measure of support for that hypothesis to "evidence" which is of the nature of a purely logical truth (e.g., the truth that the ship is either a coal-burner or not a coal-burner)—that is, we would attribute no numerical degree of probative force to an argument for a factual hypothesis, when the premises cite no empirical evidence for it. Similarly, it seems plausible to say, within the framework of established habits of inductive inference, that the mortality rate of man is better supported by the present biological evidence than is the hypothesis of telekinesis by the available parapsychological data. But I do not think that the tacit rules embodied in those established habits enable us to compare the support given by available evidence for the assumption that the earth was once part of the sun, with the support given by extant evidence for the hypothesis that Richelieu once loved Anne of Austria. In short, actual estimates of evidential weight are in general not quantitative, they are not made with respect to every consistent set of statements which may conceivably be introduced as evidential premises, and they are not always comparable.

Now Carnap's system of inductive logic not only aims to analyze and to bring into a coherent order inductive arguments which are commonly considered to be sound. His system also seeks to generalize the principles underlying those arguments, so as to bring within the scope of the broadened principles questions of inductive inference upon which common practice is usually silent. Carnap believes, in particular, that quantitative determinations of evidential support are not beyond the bounds of possibility; and he in fact devotes his major effort to the construction of systems of quantitative inductive logic.[2] There are therefore some obvious respects in which Carnap's proposed "rational reconstruction" of induction deviates from actual inductive practice.

The mere circumstance that such differences exist does not, of course, constitute a difficulty for his system. Customary standards of inductive reasoning are not beyond criticism and correction; and just as recent statistical theory has developed improved methods for the conduct of inductive inference, so Carnap's theory may also indicate ways for improving habitual modes of assessing inductive arguments. Moreover, no serious objection to his theory can be based on the fact that there are differences between his theory and customary conceptions, if the deviations occur at points at which habitual notions are vague or noncommittal, or if the innovations are simply "auxiliary" devices (introduced perhaps for the sake of achieving a uniform and formally satisfactory method for analyzing inductive arguments) which do not enter constitutively into the final assessment of the evidence for a given hypothesis.

The picture is radically altered, however, if Carnap's approach requires the adoption of methods for weighing evidence that appear to be incompatible with ostensibly reliable inductive procedures, or if his methods rest on stipulations that are either question-begging or practically unrealizable. In the former case, some proof is required that the proposed innovations are more effective for achieving the objectives of empirical inquiry than are the customary methods; in the latter case, little if anything has been accomplished in the way of a viable theory of inductive inference. It is perhaps arguable, for example, that the assumption in Carnap's theory, according to which a hypothesis receives a measurable degree of support from any consistent set of empirical premises, is in general only an innocent formal requirement, and though it deviates from established procedures the assumption has only a negligible import for the normal task of assessing empirical evidence. On the other hand, the further assumption in Carnap's system, that even purely logical truths lend a degree of support to empirical hypotheses, does not appear to be quite so innocuous. For as it turns out, this *a priori* degree of support for a hypothesis enters fundamentally into the determination of the degree of support that *empirical* evidence gives to the hypothesis. The assumption seems therefore to be entirely in disaccord with the way in which the weight of evidence is normally assessed. There are, moreover, good reasons for maintaining that modern science has achieved its successes in part because it has rejected the mode of evaluating its hypotheses which is based on that assumption. This feature of Carnap's inductive logic, taken by itself, does not necessarily deprive the system of all value, and it is certainly not unacceptable to many distinguished analysts of inductive procedure. There are, however, further assumptions underlying the system that appear to me no less debatable, and in what follows I propose to examine them.

2

Carnap bases his quantitative inductive logic on a definition of the notion of the degree of confirmation (or probability$_1$) of a hypothesis h relative to (noncontradictory) evidence e—written for short as '$c(h, e)$', and even more briefly when no confusion arises as 'c'. A fundamental condition Carnap imposes upon c is that it satisfies a set of postulates, essentially the postulates usually assumed for the mathematical calculus of probability. These postulates require, among other things, that *c* be associated with a real number in the interval from 0 to 1 inclusive, for every pair of statements *h* and *e*—provided only that *e* is not self-contradictory. On the other hand, these postulates define *c* only implicitly, so that there is a nondenumerable infinity of ways in which *c* can be *explicitly* defined so as to conform with the postulates. Carnap therefore indicates how, for a certain class of specially constructed languages, explicit definitions for the *c* can be given, each definition corresponding to what he calls an "inductive method." These languages possess a relatively simple syntactical structure, adequate for formulating certain parts of scientific discourse, though not the whole of it. Carnap's problem then reduces to that of selecting from these infinitely numerous inductive methods just those (possibly just one) which promise to be adequate for actual inductive practice and which are in reasonably good agreement with our habitual (or "intuitive") notions concerning the assessment of inductive evidence.

It turns out, however, that the infinitely numerous possible definitions fall into one or the other of two classes. The *c*'s falling into the first class are functions of the number of primitive predicates in the language for which they are defined; the *c*'s belonging to the second class are not functions of this number, but depend on a parameter whose value is assigned in some other way. Carnap appears to believe, though whether he really does so is not quite certain on the basis of his published statements, that a certain *c* belonging to the former class and designated as 'c^*' is particularly appropriate as the foundation for an inductive logic which can serve to clarify, systematize, and extend actual inductive practice. I shall therefore first discuss c^*, and postpone comment on other definitions in Carnap's repertory of inductive methods.

Every language for which *c* (and therefore c^*) is defined has a finite number of primitive predicates, and a finite or denumerably infinite number of individual constants. Although the predicates may be of any degree, Carnap develops his system in detail mainly for the case that the predicates are all monadic. Moreover, although the individuals named by the individual constants may be of any sort (e.g., physical objects, events, etc.), he suggests that for technical reasons it is preferable to take them to be spatio-temporal positions.[3] In any case, the only characteristics that are to be ascribed to the individuals mentioned in a given language are those expressible in terms of its primitive predicates and the explicit definitions constructed out of these. Accordingly, there is one indispensable condition which the primitive predicates must satisfy, if the language in which they occur and the inductive logic based on c^* are to be adequate for the aims of science: the set of primitives must be *complete*, in the sense that they must suffice to express every "qualitative attribute" we may ever have the occasion to predicate of the individuals in our universe.

The reason for this requirement of completeness is that c^* is so defined that its numerical value for a given h and e is in general a function of the number of primitive predicates in the language. Thus, suppose a language is adopted containing π independent monadic primitive predicates. Then there will be k ($= 2^\pi$) "narrowest classes" specifiable with the help of these predicates and their negations. Suppose, moreover, that 'M' is a predicate which is expressible as the disjunct of w of these narrowest classes, w being the "logical width" of M. If now the evidence e asserts that in a sample of s individuals, s_1 have the property M, and h is the hypothesis that an individual not mentioned in e also has the property M, then

$$c^*(h,e) = \frac{s_1 + w}{s + k}.$$

However, if the language is not complete, and if new primitive predicates must be added to cxpress some feature of the universe, the logical width of M in the new language will be increased, even though the *relative* logical width w/k of M will be unaltered. It follows immediately that $c^*(h,e)$ in the first language will differ from $c^*(h,e)$ in the second enlarged language. To be sure, as Carnap has explicitly noted, the values of the c^*'s will remain in the interval with the endpoints s_1/s and w/k, where s_1/s is the observed relative frequency of M in the sample of size s and w/k is the constant relative width of M. He has also pointed out that if the sample size s is increased but the relative frequency s_1/s of M remains the same, then even though the number of primitive predicates is augmented, the relative frequency s_1/s will swamp the influence of π (the number of primitive predicates) upon the value of c^*, and that as s increases without limit $c^*(h,e)$ will approach s_1/s as the limit. Nevertheless, the fact that c^* varies at all with the number of primitive predicates in the language appears to be strongly counterintuitive. Certainly no biologist, for example, would be inclined to alter his estimate of the support given by the available evidence to the hypothesis that the next crow to be hatched will be black, merely because the language of science becomes enriched through the introduction in some branch of sociology of a new primitive predicate. Nor is there any *prima facie* good reason why such an altered estimate should be made. On the other hand, if the set of primitive predicates is complete, their number cannot be augmented, and the difficulty disappears.

But is the proposed cure an improvement on the disease? Unless we do have good reasons for fixing the number of primitive predicates in a complete set, we cannot, even in principle, calculate the value of c^* for nontrivial cases, so that the inductive logic based on c^* is simply inapplicable. But the assumption that a complete set of primitives contains a given number π of predicates is not a truth of logic; it is at best a logically contingent hypothesis which can be accepted only on the basis of empirical evidence. The assumption is not a logical truth, for it in effect asserts that the universe exhibits exactly π elementary and irreducible qualitative traits, into which all other traits found in nature are analyzable without remainder. It is an assumption which would be contradicted by the discovery of some hitherto unnoted property of things (e.g., an odor or distinct type of physical force) that is not explicitly analyzable in terms of the assumed set of basic traits. Since the assumption must therefore be evaluated in the light of available empirical evidence, the obvious question arises as to how the weight of this evidence is to be estimated. It cannot be measured by way of c^* defined for the language with π primitive predicates. For in that language the assumption is an analytic truth, and its c^* has the maximum value of 1, contrary to the supposition that the assumption is a contingent hypothesis. Nor can the weight of the evidence for the assumption be measured in terms of a c^* defined for some different language. For this latter language would then have to have a complete set of primitive predicates, and we would thus be faced with an infinite regress. Perhaps a c, different from c^*, is needed, one for which the condition of completeness is not essential? But if so, there are no clues as to which alternative to c^* is to be employed; and in any event, if a c different from c^* is required in order to select a language in which c^* is to be defined, then c^* would not be the *uniquely* and *universally* adequate measure of evidential support—contrary to the supposition underlying the present discussion that c^* is such a measure.

However this may be, it is difficult to avoid the conclusion that the assumption that we have, or some day shall have, a complete set of primitive predicates is thoroughly unrealistic, and that in consequence an inductive logic based on that assumption is a form of science fiction. Although in certain areas of experience we are fairly confident that all the directly observable traits have already been noted, there are no good reasons for believing that we have already catalogued such traits occurring in all parts of the universe. All possible experiments upon all individuals spread through time have not been, and are not likely to be, performed; and the ancient discovery of the previously unknown magnetic property of loadstones has had its analogue frequently repeated in the past and may continue to be repeated in the future.

Moreover, though this point perhaps bears only on eventual developments of Carnap's system so as to make it potentially applicable to the whole of the language of science and not only to a fragment of it as is the case at present, even a presumptively complete catalogue of predicates referring to *directly observable* traits would not exhaust the primitive predicates actually required in *theoretical science*. The theoretical predicates which enter into modern systems of natural science (e.g., such predicates as "entropy," "gene," or "electron") are not explicitly definable in terms of directly observable things, though without them scientific research as we know it would be impossible. Such theoretical predicates are usually the products of great feats of scientific imagination; and the introduction of new theoretical predicates into a branch of science is often accompanied by the elimination of older ones—this has been the fate of such terms as "phlogiston" and "caloric." The theoretical parts of the language of science, at any rate, undergo frequent changes, and the direction of change does not appear to be converging towards a limit. The supposition that some day we shall have a complete list of theoretical predicates is thus tantamount to the assumption that after a certain date, no further intellectual revolutions in science will occur. But this is an assumption that is incredible on the available evidence.

As Carnap recognizes, the requirement of completeness is related to John Maynard Keynes's principle of limited variety (and incidentally, also to Francis Bacon's doctrine of "forms"), according to which the amount of variety in the universe is so limited that no one object possesses an infinite number of independent properties. Carnap does not find this principle implausible, and cites in its support the success of modern physics in "reducing" the great variety of phenomena to a small number of fundamental theoretical magnitudes.[4] But this evidence does not seem to be compelling, if only because it is at least debatable whether the phenomenal qualities of things are *explicitly definable* in terms of the theoretical concepts of physics; and if they are not definable, the total number of primitive predicates has not in fact been diminished. Moreover, though no *a priori* limits can be set to the scope of physical theory, and it may well be that the physics of the future will account for larger areas of our phenomenal experience than it does at present, two points should be noted. In the first place, current physical theory does not in fact embrace all that experience, and it may never do so. In the second place, the evidence of history seems to show that as the scope of physics is enlarged, the number of its primitive theoretical predicates does not converge to any fixed value, and no plausible upper bound can be assigned to such a number, if indeed there is one. But without a reliable estimate of the value of such an upper bound (to say nothing of offering a reasonably based conjecture as to what will be the actual primitive predicates that a possibly complete physical theory of the future will require), a fully satisfactory inductive logic based on c^* cannot be constructed. The fulfillment of the requirement of completeness depends on our possessing more knowledge than we possess at present, or are likely to possess in the foreseeable future. And if the requirement should ever be fulfilled, we would, by hypothesis, have acquired so much knowledge about the universe that much of our present need for an inductive logic will no longer be actual.

3

There is a further difficulty (which may, however, be only an apparent one) that faces an inductive logic based on c^*—and more generally, on a c that is a function of the number of primitive predicates in the language. It is a familiar fact that two deductive systems may be logically equivalent, so that statements in one are translatable into statements in the other and conversely, even though each system is based on a distinctive set of primitive predicates and a distinct set of axioms. Thus, Euclidean geometry can be developed in the manner of Veblen (who employs, among others, the terms "point" and "between" as primitives), or in the fashion of Huntington (who uses "sphere" and "includes" as primitive predicates); and there is no statement in the Veblen system which cannot be matched in the Huntington codification, and vice versa. If two languages, each containing only monadic primitive predicates, are intertranslatable, then it can be shown that the number of primitives in one must be equal to the number in the other. But in general, if at least one of two intertranslatable languages has polyadic primitives then it seems that the number of primitives in one may be different from the number in the other. But if this is so, the consequences are serious. For suppose that a hypothesis *h* and the evidence *e* for it can be formulated in two intertranslatable languages L_1 and L_2, where the number of polyadic primitives in the former is unequal to the number of primitives in the latter. It then follows that since c^* is a function of the number of primitive predicates in the language for which it is defined, the value of $c^*(h,e)$ calculated for L_1 will be unequal to the value of $c^*(h,e)$ calculated for L_2. Accordingly, the degree of support which the same evidence provides for a given hypothesis will depend on which of two equivalent languages is used for codifying the evidence and the hypothesis. This result is strongly counterintuitive. If the premise of this argument is sound (and I frankly do not know whether it is or not), then the degree of support which a hypothesis receives from given evidence on the basis of Carnap's approach is contingent on the arbitrary choice of one among several logically equivalent languages. But such a conception of evidential weight seems of dubious value as the basis for the practice of induction.

Two considerations occur to me, however, which may make this difficulty only a spurious one. One of them is Carnap's suggestion that in addition to satisfying the requirement of completeness, the primitive predicates of a language for which c^* is defined must also satisfy the requirement of simplicity. As Carnap once formulated this requirement, "the qualities and relations designated by the primitive predicates must not be analyzable into simpler components."[5] The required simplicity of primitive predicates must, on this stipulation, be an "absolute" one, and not merely relative to some given language or mode of analysis. If this notion of simplicity could be assumed to be sufficiently clear, the difficulty under discussion would presumably vanish. For if two intertranslatable languages are constructed on the basis of two sets of unequally numerous primitive predicates, it might be possible to show in general that one of the sets of primitives is simpler than the other, and that therefore the value of c^* must be calculated for the simpler of the two languages. Nevertheless, the notion of absolute simplicity is far from clear. If we do not employ psychological criteria such as familiarity, what rules are to be used in deciding whether, for example, the Veblen set of primitives for geometry are simpler than the Huntington set? When Carnap first proposed the requirement of simplicity he himself admitted his inability to give an exact explication of the notion; and the obscurity of the notion perhaps accounts for the fact that he has not mentioned this requirement in more recent publications. But in any event, the use of the notion of absolute simplicity for outflanking the above difficulty in c^* generates difficulties that are no less grave.

The second consideration mentioned above is of a more technical sort. The value of c^* is in fact an explicit function of the number of *state-descriptions* constructable in the language adopted, and only indirectly of the number of primitive predicates in it. Now a state-description states for every individual, and for every property or relation designated by the primitives, whether or not the individual has the property or relation. Accordingly, a state-description is a noncontradictory conjunction of atomic statements or of their negations (or, in case of languages with an infinity of individual constants, it is an infinite class of such statements)—where an atomic statement ascribes a property (or relation) designated by a primitive predicate to an individual (or individuals) named by an individual constant (or constants). It follows that if the primitive predicates of a language are not totally logically independent of each other, not every conjunction of atomic statements or of their negations will be a state-description—since in that case some of the conjunctions will be self-contradictory in virtue of the logical relations between the predicates, and must therefore be omitted from the count of all possible state-descriptions. But if two intertranslatable languages L_1 and L_2 are based on two unequally numerous sets of promitive predicates, there will presumably be relations of logical dependence between the primitives in each set, so that the number of state-descriptions in L_1 will be the same as the number of state-descriptions in L_2. It will then follow that for given h and e, the values of $c^*(h,e)$ will also be the same for the two languages, so that the objection to c^* under present discussion loses its point.

There are, however, two comments on this solution of the difficulty which seem to me in order. The solution assumes that it is possible to give, for each of two intertranslatable languages, an exhaustive catalogue of the rules or postulates which specify the relations of logical dependence between its primitives. Such a catalogue can of course be offered for "artificial" languages, since artificial languages are actually constructed by explicitly stipulating what are the relations of logical dependence between the primitives and what are the logically contingent connections between them. But this is not so readily accomplished for the so-called "natural" languages (including much of the language of science), for in such languages it is not always clear which statements are logically necessary and which have the status of logically contingent hypotheses. Indeed, as is well-known, the same *sentence* may alter its status in this respect with the progress of inquiry or with alternate codifications of a scientific system. (For example, the sentence expressing the ostensibly contingent second law of motion in Newton's system of mechanics, appears as a statement of a logical truth in Mach's reformulation of the system.) On the other hand, though this problem of codifying a natural language is in practice often a difficult one, it is not a problem that is distinctive to Carnap's system of inductive logic.

The second comment is this. Carnap has thus far defined the notion of logical width only for languages with monadic predicates. But it seems plausible to assume that when he does develop his system for more complex languages, he will require an analogous notion for the latter. I have no idea how he will define the notion for the general case. However, it seems to me a reasonable conjecture that for languages with polyadic predicates, as for languages with exclusively monadic ones, the logical width of a predicate must also be some function of the number of primitive predicates in the system. But if this conjecture is sound, an important question immediately arises, one which bears directly on the adequacy of the suggested solution to the difficulty under discussion. Given two intertranslatable languages based on two sets of unequally numerous primitives, and granted that the number of state-descriptions in each is the same, will corresponding predicates in the two languages (i.e., predicates that designate the same property) also have the same logical width? If not, and since the logical width presumably enters into the value of c^*, then for given h and e the value of $c^*(h,e)$ in the two languages will not be the same. In that eventuality, however, the difficulty under discussion will not have been put to final rest.

4

I wish next to raise an issue that concerns not only c^* but also the whole continuum of inductive methods Carnap regards as possible candidates for explicating the notion of evidential support. Among the conditions he lays down which any reasonable c must satisfy, there are two that bear considerable resemblance to the notorious Principle of Indifference, often regarded as the Achilles heel of the classical theory of probability. The first of these stipulates that all the individuals are to be treated on par, the second introduces a similar requirement for primitive predicates. According to the first, for example, if the evidence e asserts that the individuals a_1 and a_2 have the property M, while the hypothesis h declares that the individual a_3 has M, then $c(h,e)$ must be equal to $c(h',e')$, where e' asserts that the individuals a_4 and a_5 have M, and h' declares that the individual a_6 has M. According to the second requirement, if 'P_1' and 'P_2' are primitive predicates, e asserts that a_1 and a_2 have the property P_1, and h asserts that a_3 has P_1, then $c(h,e)$ must equal $c(h',e')$—where e' declares that a_1 and a_2 have the property P_2 and h' declares that a_3 has P_2. In short, $c(h,e)$ must be invariant with respect to any permutation of individual constants as well as with respect to any permutation of the primitive predicates.

These requirements are initially plausible, and as Carnap points out assumptions very much like them are tacitly employed in deductive logic. Taken in context, they formulate a feature of actual inductive practice; and in generalizing sciences like physics they are unavoidable, on pain of putting an end to the use of repeated experiments for establishing universal hypotheses. For example, it obviously makes no difference to the evaluation of the evidence for the generalization that water expands on freezing, whether the evidence is obtained from one sample lot of water rather than another sample—*provided* that the samples are taken from a reservoir of the substance that is homogeneous in certain respects. Similarly, it makes no difference to the credibility of a generalization, whether the generalization under inquiry is that copper expands on heating or whether the generalization is that copper is a good electrical conductor—*provided* again that the instances used as evidence are the same in both cases, and *provided* also that the hypothetical relations between the properties under investigation are assumed to be dependent only on the traits of things explicitly mentioned.

On the other hand, as the examples just mentioned suggest, such judgments of indifference are made within contexts controlled by *empirical assumptions* as to what are the relevant properties of individuals that must be noted in using the individuals for evidential purposes, and as to the relevant factors that must be introduced into general statements concerning the concomitances of properties. Thus, different samples of water must be sufficiently homogeneous in their chemical composition, though not in their historical origins, if they are to be on par as evidence for the generalization concerning the expansion of water when cooled. Again, if given instantial evidence is to carry the same weight for the generalization that copper expands when heated as it does for the generalization that copper is a good conductor, the concomitances asserted must be assumed to be independent of variations in other properties exhibited by the instances—for example, of differences in the shapes or the weights of the individuals upon which observation is being made. It is clear, however, that these judgments of relevance and irrelevance are based on *prior experience* and cannot be justified by purely *a priori* reasoning.

Within the framework of Carnap's construction, however, the status of the requirements concerning the indicated invariance of *c* is different. For in his system, the invariance is absolute, not relative to special contexts involving special empirical assumptions. Indeed, the invariance is postulated antecedently to any empirical evidence which might make the postulation a reasonably plausible one. It is not easy to see, therefore, what grounds—other than purely arbitrary and *a priori* ones—can be adduced for such a requirement of absolute invariance. Carnap defends the requirement by arguing in effect that since the primitive predicates are stipulated to be logically independent, there is no reason for assigning unequal *c*'s to two hypotheses relative to the evidence for them, when the respective hypotheses and evidential statements are isomorphic under a permutation of individual constraints or primitive predicates. But though there is no *a priori* reason for assigning unequal c's in such a case, neither is there a compelling *a priori* reason for assigning *equal* ones. There is surely the alternative, suggested by actual scientific practice, that the matter is not to be decided once for all by fiat, but settled differently for different classes of cases in the light of availble empirical knowledge. In any event, the value of the Principle of Indifference is as debatable (when it is used, as Carnap uses it, to specify in inductive logic which pairs of hypotheses and evidential statements are significantly isomorphic) as when the principle is employed in the classical manner to determine the magnitudes of empirical probabilities.

5

Some of the consequences which follow from the adoption of c^* as the measure of evidential support must now be examined. One of these consequences is that for a language with an infinity of individual constants (i.e., in a universe with a nonfinite number of individuals), the value of c^* for any universal empirical hypothesis, relative to any finite number of confirming instances for it, is always zero. For example, despite the great number of known corroborative instances for the generlaization that water expands on freezing, the evidential support provided by those instances for this ostensible law is zero when measured by c^* in an infinite universe, and is no better than the evidential support given by those instances to the contrary hypothesis that water contracts on freezing. Moreover, even if the number of individuals in the universe is assumed to be finite but very large, the c^* for the generalization relative to the available instantial evidence will normally differ from zero only by a negligible amount.

Accordingly, if c^* were a proper measure of what ought to be our degree of reasonable belief in hypotheses, none of the generalizations proclaimed by various sciences as laws of nature merits our rational confidence. The search by scientists for critical evidence to support such claims is then pointless, for however much evidence is accumulated in favor of universal laws, increments in the degree of that support remain at best inappreciable. This outcome of adopting c^*, however, is patently in disharmony with our customary way of judging such matters.

There are several ways, nevertheless, in which this apparently fatal consequence entailed by c^* may be made more palatable. In might be argued, in the first place, that it is gratuitous to assume the universe to contain an infinity of individuals, so that the theorem concerning the value of c^* for universal hypotheses in infinite languages simply does not apply to the actual world. It must of course be admitted that we have no certain knowledge that our universe does indeed contain an infinity of empirically specifiable individuals, even if the universe is taken to be extended in time without limit. On the other hand, neither do we know with certainty that the individuals in the universe are only finite in number. If the use of c^* is defensible only on condition that this number really is finite, its use must be postponed indefinitely until that fact is established; and we shall have to carry on our inductive studies (including the inquiry into the number of individuals in the universe) without the help of an inductive logic based on c^*. Moreover, as has already been noted, even if the universe contains only a very large finite number of individuals—and there surely is competent evidence that this number is very large indeed—for all practical purposes such a number entails the same undesirable consequences as if it were nonfinite.

It might be claimed, in the second place, that it is just a mistake to raise questions about the "probability" of universal hypotheses, and thereby a mistake to view them as statements on par with instantial ones, concerning which it is significant to ask what measure of support they receive from given evidence. For universal hypotheses, so it is often said, function as guides to the conduct of inquiry, as instruments for predicting concrete events, or as means for organizing systematically the outcome of previous investigations. Universal hypotheses, on this view, are intellectual devices concerning which it is appropriate to ask whether they are adequate for achieving the ends for which they have been designed, but not whether they are true or false. Accordingly, the circumstance that for universal hypotheses the value of c^* is uniformly zero, simply calls attention to the absurdity of treating them as factual statements for which evidence is to be assessed. However, whatever the merits or limitations of this intellectual gambit may be, it is not one which Carnap can employ. For on his approach, universal hypotheses are considered to be on par with instantial ones in respect to their status as empirical statements. It is indeed a central feature of his system that for *any* hypothesis h and (noncontradictory) evidence e, the degree of confirmation $c^*(h, e)$ must have a determinate value.

Carnap's own proposed resolution of the difficulty bears a certain resemblance to the one just mentioned. But he offers a technically different answer, by way of the notion of the "instance confirmation" of universal laws. He introduces his discussion (though a full account by him is still not available) with the following general explanation:

> Suppose we ask an engineer who is building a bridge why he has chosen the particular design. He will refer to certain physical laws and tell us that he regards them as 'very reliable,' 'well founded,' 'amply confirmed by numerous experiences.' What do these phrases mean? It is clear that they are intended to say something about probability$_1$ or degree of confirmation. Hence, what is meant could be formulated more explicitly in a statement of the form '$c(h, e)$ is high' or the like. Here the evidence e is obviously the relevant observational knowledge. But what is to serve as the hypothesis h? One might perhaps think at first that h is the law in question, hence a universal sentence l of the form: 'For every space-time point x, if such and such conditions are fulfilled at x, then such and such is the case at x.' I think, however, that the engineer is chiefly interested not in this sentence l, which speaks about an immense number, perhaps an infinite number, of instances dispersed through all time and space, but rather in one instance of l or a relatively small number of instances. When he says that the law is very reliable, he does not mean to say that he is willing to bet that among the billion of billions, or an infinite number, of instances to which the law applies there is not one counterinstance, but merely that this bridge will not be a counterinstance, or that among all bridges which he will construct during his lifetime there will be no counterinstance. Thus h is not the law l itself but only a prediction concerning one instance or a relatively small number of instances. Therefore, what is vaguely called the reliability of a law is measured not by the degree of confirmation of the law itself but by that of one or several instances.[6]

Carnap thereupon defines the instance confirmation of a law l on evidence e as the c^* value of the support given by e for the hypothesis that an individual not mentioned in e fulfills l. Furthermore, he defines the qualified-instance confirmation of the law l as the c^* value of the support given by e to the hypothesis that an individual not mentioned in e, but possessing the property mentioned in the antecedent clause of the universal conditional l, also has the property mentioned in the consequent clause of l.

Carnap then argues that contrary to usual opinion, the use of laws is not essential for making predictions, since the inference to a new case can be made *directly* from the instantial evidence, rather than through the mediating office of the law. Thus, suppose the hypothesis h under discussion is whether some given individual a has the property B, and that the evidence e asserts that all of the many other individuals which have been observed to possess the property A also possess B. Suppose further that j is the instantial datum that a has A. The usual account, as Carnap formulates it, is that from e we first inductively infer the law l: All A's are B's; and from l together with j we deductively infer h. However, since $c^*(l,e)$ is zero or very close to it, this argument is unsatisfactory. But according to Carnap we really do not need a high value for $c^*(l,e)$ in order to obtain a high value for $c^*(h,e.j)$—that is, for a qualified-instance confirmation of the law. In his view, on the contrary, the person X conducting the inquiry

> need not take the roundabout way through the law l at all, as is usually believed; he can instead go from his observational knowledge $e.j$ directly to the singular prediction h. That is to say, our inductive logic makes it possible to determine c^* $(h,e.j)$ directly and to find that it has a high value, without making use of any law. Customary thinking in everyday life likewise often takes this short cut, which is now justified by inductive logic. For instance, suppose somebody asks X what he expects to be the color of the next swan he will see. Then X may reason like this: he has seen many white swans and no nonwhite swans; therefore he presumes, admittedly not with certainty, that the next swan will likewise be white; and he is willing to bet on it. Perhaps he does not even consider the question whether all swans in the universe without a single exception are white; and, if he did, he would not be willing to bet on the affirmative answer.
>
> We see that the use of laws is not indispensable for making predictions. Nevertheless it is expedient, of course, to state universal laws in books on physics, biology, psychology, etc. Although these laws stated by scientists do not have a high degree of confirmation, they have a high qualified-instance confirmation and thus serve as efficient instruments for finding those highly confirmed singular predictions which are needed in practical life.[7]

In short, Carnap appears to be in substantial agreement with J. S. Mill's view that the fundamental type of inductive reasoning is "from particulars to particulars."

Carnap's proposed solution of the difficulty is brilliantly ingenious. But is it satisfactory? Several considerations make this doubtful to me:

i) His solution is predicated on the assumption that the evidence in the qualified-instance confirmation of a law can in general be identified and established without even the tacit acceptance and use of laws, since otherwise a regress would be generated that would defeat the object of his analysis. The assumption is illustrated by his own example, in which the instantial statement '*a* is a swan,' constituting part of the evidence for the hypothesis that *a* is white, can presumably be affirmed on the strength of a direct observation of the individual *a* without the implicit use of any universal laws. I shall not dispute this particular claim, even though legitimate doubts may be expressed as to whether the assertion that *a* is a swan does not "go beyond" what is directly present to observation, and does not carry with it implicit assumptions about invariable connections between anatomical structure, physiological function, and other biological properties—connections which are assumed when an organism is characterized as a swan. But I do dispute the ostensible claim that this example is typical of the way laws are in general confirmed, and that the instances which confirm many scientific theories are quite so simply obtained. To be sure, most of these theories cannot be formulated in the restricted languages for which Carnap has thus far constructed his system of inductive logic; but I do not believe that point under discussion is affected by this fact. Consider, therefore, some of the confirming instances for the Newtonian theories of mechanics and gravitation. One of them is the obloid shape of the earth. The fact that it is obloid, however, can be established only through the use of a system of geometry and of optical instruments for making geodetic measurements—all of which involve at least the tacit acceptance of universal laws as well-founded. Could we inductively infer this fact from the instantial evidence alone, without including in the evidence for it any general laws? It would not advance the solution of the problem, were we to construct an inductive argument, to parallel the schema suggested by Carnap, to read as follows: The qualified-instance confirmation of the Newtonian laws, where the instance is the obloid shape of the earth, is high relative to the instantial evidence that all of the many rotating solids which have been observed in the past have an equatorial bulge, supplemented by the additional evidence that the earth is a rotating solid. For how can the fact that the earth is a rotating solid be established, except by way of assuming an astronomical theory? But unless this fact (or an analogous one) is granted, it is difficult to see how Newtonian theory is relevant to ascertaining the earth's shape, or to understand why the earth's obloid shape is to be counted as a confirming instance for that theory. I do not believe, therefore, that Carnap has successfully defended c^* against the objection that it leads to grave difficulties when it is applied to universal hypotheses.

ii) There is a further point bearing on the present issue, which is suggested by Carnap's discussion of Laplace's Rule of Succession. As is well known, Laplace derived a theorem from the assumptions of his theory of probability, which asserts that if a property is known to be present in each member of a sample of s events, the probability that the next event will also have this property is

$$\frac{s+1}{s+2}.$$

Using the evidence available to him concerning the past risings of the sun, Laplace then calculated the probability of another sunrise to be

$$\frac{1{,}826{,}214}{1{,}826{,}215}.$$

This result has been severely criticized by a number of authors for a variety of reasons. Carnap also finds Laplace's conclusion unsatisfactory, because Laplace allegedly violated the "requirement of total evidence." According to this requirement, "the total evidence available must be taken as basis for determining the degree of confirmation" [probability$_1$] in the application of the theorems of inductive logic to actual situations.[8] Carnap points out that Laplace assumed the available evidence to consist merely of the known past sunrises, and that he thereby neglected other evidence for the hypothesis that the sun would rise again—in particular, the evidence involved in his knowledge of mechanics. As Carnap puts the matter,

> the requirement of total evidence is here violated because there are many other known facts which are relevant for the probability of the sun's rising tomorrow. Among them are all those facts which function as confirming instances for the laws of mechanics. They are relevant because the prediction of the sunrise for tomorrow is a prediction of an instance of these laws.[9]

Although I do not believe, as Carnap does, that there is no analogue to the requirement of total evidence in deductive logic,[10] I shall not pursue this side issue, and will assume that Carnap's diagnosis of Laplace's error is well taken. The question I do wish to raise is whether it is the *confirming instances* of the laws of mechanics, or the *laws of mechanics* themselves, which are to be included in the evidence when tomorrow's sunrise is predicted. Carnap appears to adopt the former alternative. It is not clear, however, why in that case most of the evidence—taken simply as so many *independent* instantial statements—is *relevant* to the prediction of another sunrise, and why it should raise the c^* for the predictive hypothesis. For example, the confirming instances of the laws of mechanics include observations on tidal behavior, on the motions of double stars, on the rise of liquids in thin tubes, on the shapes of rotating liquids, and much else, in addition to observations on the rising of the sun. There is, however, no purely logical dependence between instantial statements about the height of the tides or instantial statements on the rising of the sun on the other hand. Apart from the laws of mechanics, these statements express just so many disparate facts, no more related to each other than they are related to other statements which do *not* formulate confirming instances of these laws—for example, statements about the magnetic properties of a given metal bar, or about the color of a man's eyes. Why should the inclusion of the former instantial data increase the evidential support for the prediction of another sunrise, but the inclusion of the latter ones not do so?

To make the point more fully, consider a language with three logically independent monadic primitive predicates 'P_1', 'P_2' and 'P_3'. Then according to the definition of c^*,

$$c^*(P_1a_1, P_1a_2) = 5/9$$

and if the evidence is enlarged to include P_1a_3,

$$c^*(P_1a_1, P_1a_2.P_1a_3) = 6/10,$$

so that the c^* for the hypothesis is increased. But if the evidence is further enlarged by including P_2a_4, then

$$c^*(P_1a, P_1a_2.P_1a_3.P_2a_4) = 6/10,$$

so that this additional evidence is irrelevant; and the situation remains the same when the evidence is further augmented by adding to it P_3a_5, or P_2a_6, or in fact any number of instantial statements which ascribe properties to individuals other than the property designated by the predicate 'P_1'. Now most of the predicates occurring in the formulation of confirming instances for the laws of mechanics are *prima facie* quite analogous to the predictive predicates in this example in respect to their being logically independent of each other. Indeed, there are cases in the history of science when, on the basis of some well established theory, an event has been predicted which had rarely if ever been observed previously. In such cases, though the predictive hypothesis receives a considerable measure of support from the theory, the predicates in the instantial evidence for the theory are for the most part different from, and logically independent of, the predicates occurring in the hypothesis. To cite a notorious example, William R. Hamilton predicted from theoretical considerations, the phenomenon of conical refraction, though this phenomenon had not been previously observed, so that instances of the phenomenon did not constitute a part of the evidence for the theory Hamilton employed. In consequence of all this, it does not appear plausible that, in conformity with the requirement of total evidence, it is the *instantial evidence* for the laws of mechanics, but rather the *laws of mechanics* themselves, which must be included in the evidence for the hypothesis of another sunrise.

But if this is so, I am also compelled to conclude that the use of general laws in inductive inference is not eliminable, in the manner proposed by Carnap. Accordingly, the notion of instance confirmation (or qualified-instance confirmation) of a law as a measure of the law's "reliability" does not achieve what he thinks this notion can accomplish. In short, I do not believe he has succeeded in outflanking the difficulty which arises from the counterintuitive consequences of adopting c^* as a measure of evidential support in infinite languages.

6

I must now raise an issue that affects not only an inductive logic based on c^*, but nearly all of the inductive methods Carnap has outlined.

It is commonly assumed that the evidential support for a hypothesis (whether singular or universal) is generally increased by increasing the sheer number of its confirming instances. For example, it is usually claimed that the hypothesis that the next marble to be drawn from an urn will be white is better support by evidence consisting of

100 previous drawings each of which yielded a white marble, than by evidence consisting of only 50 such drawings; and many accounts of inductive logic attribute this difference in evidential weight entirely to the difference in the relative size of the two samples. Again, it is often supposed that simply by repeating an experiment on the period of a pendulum, where each experiment shows this period to be proportional to the square-root of the pendulum's length, the weight of the evidence for the generalization that the period of any pendulum follows this law is augmented. In any event, this assumption is implicit in most of the inductive methods (including the one based on c^*) which Carnap discusses. But although the assumption appears to be eminently plausible, I think it is a reasonable one only when it is employed under certain conditions, so that the assumption is acceptable only in a qualified form. I want to show, however, that most of Carnap's inductive methods in effect adopt it without such qualifications.

Consider first a language with two monadic primitive predicates 'R' and 'S' and N individual constants, so that if 'Q' is defined as 'R.S,' 'Q' specifies one of the four narrowest classes of individuals which is formulable in this language. The relative logical width of 'Q' is then $w/k = 1/4$. Suppose now that a sample of size s is drawn from the population, that just s_i individuals in the sample have the property Q, and that h is the hypothesis asserting that an individual a not contained in the sample also has Q. Carnap shows that for all the measures c satisfying the conditions he regards as minimal for a measure of evidential support,

$$c(h, Qa_1 \ldots Qa_{s_i} \cdot —Qa_{s\,i+1} \ldots —Qa_s) = \frac{s_i + \lambda/4}{s + \lambda}.$$

with $0 \leq \lambda \leq \infty$, where the value of the parameter λ depends on the inductive method adopted and thus fixes the measure c of evidential support. For c^*, this parameter is equal to k (the total number of narrowest classes specifiable in the language), and in the present example is equal to 4. When all the individuals in the sample have the property Q, $s_i = s$ and

$$c(h, Qa_1 \ldots Qa_s) = \frac{s + \lambda/4}{s + \lambda} = \frac{4s + \lambda}{4(s + \lambda)}.$$

Suppose now the size of the sample is increased by n, so that it contains $s + n$ individuals, and that every member of the sample has Q. Then

$$c(h, Qa_1 \ldots Qa_{s+n}) = \frac{4(s+n)+\lambda}{4(s+n+\lambda)}.$$

Since for $\lambda > 0$ and $n > 0$,

$$\frac{4(s+n)+\lambda}{4(s+n+\lambda)} > \frac{4s+\lambda}{4(s+\lambda)},$$

it follows that

$$c(h, Qa_1 \ldots Qa_{s+n}) > c(h, Qa_1 \ldots Qa_s).$$

Accordingly, when all the individuals in a sample belong to the class Q (so that, since Q determines one of the narrowest classes specifiable in the language, the individuals are indistinguishable in respect to the properties they exhibit) and if the sample size is increased, the measure of evidential support for the hypothesis is also increased. Indeed, in an infinite language if all the individuals in progressively more inclusive samples belong to Q and if the sample size is increased without limit, the degree of confirmation for the hypothesis approaches the maximum value of 1.

Suppose, next, that 'All A is B' is the formulation of a law in a language having only monadic predicates and N individual constants, where 'A' and 'B' are any molecular predicates defined in terms of the primitives, and where the predicate '$A \cdot -B$' has the logical width w. (It is clear that the law is logically equivalent to 'Nothing is $A \cdot -B$'.) Suppose, further, that the evidence e for the law asserts that s distinct individuals do not have the property $A \cdot -B$ and that all the individuals fall into *one* of the k narrowest clauses specifiable in the language. (Then, e is the conjunction of s instantial statements, each of which asserts that some individual has the property determining this class—the property in question being incompatible with $A \cdot -B$.) Carnap then shows that when s is very large in relation to k, c^* (All A is B, e) is approximately equal to $(s/N)^w$.[11] Accordingly, if the evidence for the law is increased by the addition of further instances all of which continue to fall into the same narrowest class, the degree of confirmation for the law is also increased. For infinite languages (i.e., when N = ∞) this degree of confirmation is of course zero, as has already been noted. On the other hand, if the logical width of the predicate 'A · B' is w', and if the evidence consists of a sample of s individuals all of which have the property A · B and all of which, moreover, fall into the same narrowest class, Carnap proves that for the measure c^* the qualified-instance confirmation of the law is equal to

$$1 - \frac{w}{s + w + w'}$$

This latter is the value of the degree of confirmation of the hypothesis that an individual, known to have the property A, also has the property B, on the evidence that s other individuals all falling into one of the narrowest classes have both A and B. Since the value of the qualified-instance confirmation of the law is independent of the number of individual constants in the language, it will differ from zero even for infinite languages, and it will be close to 1 when s is made sufficiently large.

In my judgment, however, these results are incongruous with the normal practice of scientific induction, as well as with any plausible rationale of controlled experimentation. For according to the formulas Carnap obtains for his system, the degree of confirmation for a hypothesis is in general increased if the confirming instances for the hypothesis are multiplied—*even when the individuals mentioned in the evidence cannot be distinguished from each other by any property expressible in the language for which the inductive logic is constructed*. But it seems to me most doubtful whether under these conditions we would in fact regard the evidential support for a hypothesis to be strengthened. Suppose we undertook to test a proposed law, say the law that all crows are black, by making a number of observations or experiments on individuals; and suppose further that the individuals we examined were *known to be completely alike* in respect to all the properties which we can formulate. Would there be any virtue in repeating the observations or experiments in such a case? Would we not be inclined to say that under the imagined circumstances *one* observation carries as much weight as an *indefinite number* of observations?

We do, of course, repeat observations and experiments intended to test proposed laws. But apart from our desire to make allowances for and to correct personal carelessness and "random" errors of observation, we do so only when we have some grounds for believing that the individuals are *not* completely alike in the properties they possess. In fact, we generally *select* the individuals upon which tests are to be performed so that they are *unlike* in as large a variety of features as possible, compatible with the requirement that the individuals exhibit the properties mentioned in the antecedent clause of the law we are testing. The rationale for this standard procedure is to show that the connections between properties asserted by the proposed law do hold in just the way the proposed law asserts them to hold, and that the hypothetical connections are not contingent upon the occurrence of some other property not mentioned by the proposed law. Accordingly, test-cases for the law that all crows are black will be drawn from a wide assortment of geographic regions, climatic conditions, and other variable circumstances under which crows may be found, in the hope that despite variations in these circumstances the color of the plumage is indeed uniformly associated with the anatomical structure that identifies crows, and in the desire to show that the color does not depend on the occurrence of some other properties which crows may have. In short, the sheer repetition of confirming instances does not, by itself, appear to carry much weight in the support given by the evidence to a hypothesis. But if this point is well taken, all those inductive methods considered by Carnap (including c^*) in which a contrary result is obtained (the only method for which such a contrary result does not hold is the one for which $\lambda = \infty$) are inadequate rational reconstructions of generally accepted canons of scientific inquiry.

The point just argued also has some bearing on the notion of the instance confirmation of a law. For if, as is required by Carnap's analysis, increasing the number of otherwise indistinguishable confirming instances for a law augments the degree of confirmation for a still unobserved additional instance of the law, why should not the degree of confirmation also be augmented by an equal amount for an *indefinite* number of further instances of the law—or even for the law itself? Under the conditions supposed, is there really a better reason for expecting that a *single* hitherto untested individual will conform to the law than for the hypothesis that many such individuals will do so? Thus, if all observed instances of crows are black, and if these instances are *known* to be alike in all respects formulable in the (hypothetically complete) language we employ (e.g., the crows observed come from the same locality, they have the same genetic constitution, their diet is the same, etc.), why should this evidence give stronger support to the prediction that the next crow to be observed will be black, than for the hypothesis that the next ten crows to be observed will be black, or for the hypothesis that all crows are black? The contrary view seems to me to reflect sound inductive practice.

To see the point more clearly, consider the following schematic example constructed in conformity with Carnap's procedure. Assume a language with four monadic primitive predicates 'P_1', 'P_2', 'P_3' and 'P_4'. The law 'Anything that is both P_1 and P_2 is also P_3' is proposed for testing. A sample consisting of 2*s* individuals is now examined, and each individual is found to possess the property $P_1 \cdot P_2 \cdot P_3$. Two possible cases will be considered: 1) All the 2*s* individuals belong to the narrowest class determined by $P_1 \cdot P_2 \cdot P_3 \cdot P_4$, and are therefore otherwise indistinguishable; ii) only *s* individuals in the sample belong to this class, while the remaining half belong to the narrowest class determined by $P_1 \cdot P_2 \cdot P_3 \cdot \sim P_4$.

Now the evidence in the first case leaves it unsettled whether P_3 is always present when the property $P_1 \cdot P_2$ alone is present, or whether the occurrence of P_3 is contingent not only on the presence of this property but also on the presence of P_4 as well. In the second case, however, the evidence shows that P_3 is dependent only on $P_1 \cdot P_2$, irrespective of the presence or absence of P_4. It therefore appears reasonable to maintain that the evidence in the second case is better than in the first as a support for the hypothesis that an individual not included in the sample, but known to possess $P_1 \cdot P_2$, also possesses P_3—and if I judge the matter aright, such a claim is in agreement with standard scientific practice. Now this point is also recognized by Carnap, since the value of c^* in the second case is higher than the value in the first case. It is clear, therefore, that the variety of instances contributes to the strength of evidence for a hypothesis. On Carnap's analysis, however, complete absence of variety in the instances is compatible with a high degree of confirmation, provided that the sheer number of instances is large; and this seems to me a defect in his system. Accordingly, if my argument holds water, his system fails to take into consideration an essential feature of sound inductive reasoning.

Moreover, I have not been able to persuade myself that the evidence in the first case supports to a lower degree the hypothesis that an *indefinite* number of further individuals possessing $P_1 \cdot P_2$ also possess P_3, than it supports the hypothesis that just one further individual is so characterized. It might be retorted that to deny this is counterintuitive—since we normally do say, for example, that on the evidence of having drawn 100 white marbles from an urn, the "probability" of getting a white marble on the next trial is greater than the "probability" of getting two white marbles on the next two trials. But I do not find this rejoinder convincing. If we *know* that the 100 white marbles constituting the evidence are fully alike *in all relevant respects* upon which obtaining a white marble from the urn depends, then it seems to me that the evidence supports the hypothesis that any further marbles, resembling in those respects the marbles already drawn, will also be white—irrespective of how many further marbles will be drawn from the urn. It is because we generally do *not* know what are the complete set of properties in respect to which marbles may differ, and because we therefore do *not* know whether the marbles in the sample lot are alike in all relevant respects, that the evidence offers better support for the hypothesis concerning a single additional marble than it does for the hypothesis concerning two or more additional marbles. I cannot therefore evade the conclusion that because of the consequences noted, c^* as the measure of evidential support runs counter to sound inductive practice.

7

I have been concerned thus far mainly with those inductive methods (chiefly with c^*) in the continuum of methods constructed by Carnap, in which the degree of confirmation depends on the number of primitive predicates in the language adopted. I wish now to discuss some of the assumptions underlying Carnap's construction of his continuum of methods, some features of those methods in which the degree of confirmation is not a function of the number of primitive predicates, and some of the considerations Carnap advances for adopting one method rather than another.

A fundamental assumption entering into the construction of Carnap's continuum of inductive methods is the following. If 'M' is a monadic predicate whose relative logical width is w/k, e_M is the evidence that in a sample of s individuals s_M possess the property M, and h_M is the hypothesis that an individual not included in the sample also possesses M, then the degree of confirmation of h_M on the evidence e_M must lie in the interval whose end-points are the relative frequency s_M/s and the relative logical width w/k. In fact, the value of $c(h_M, e_M)$ is set equal to a weighted mean of the "empirical factor" s_M/s and the "logical factor" w/s—so that eventually this value is taken to be equal to

$$\frac{s_M + \lambda(w/k)}{s + \lambda},$$

where λ is a parameter which may have any value from 0 to infinity.

But why should the values of $c(h_M, e_M)$ be assumed to fall into the interval determined by these end-points? Carnap argues, in the first place, that

> Other things (including s) being equal, those values [of c^*] are higher, the greater s_M/s. This has often been stated explicitly and may be regarded as one of the fundamental and generally accepted characteristics of inductive reasoning. Moreover, all known methods of confirmation or estimation for rf [relative frequency] agree that in any case of a sufficiently large sample the value of c or E [estimate function], respectively, is either equal or to close to s_M/s.[12]

I do not think, however, that this reasoning is entirely convincing. Carnap is doubtless correct in holding that "other things being equal," *estimates* of the relative frequency of some property in a population are often assumed to be close to the observed relative frequency in the samples drawn. But this is not always the case; and we do not, in general, measure the evidential support for a singular hypothesis by the relative frequency found to occur in a sample. Thus, in estimating a relative frequency in a population, much depends not only on the observed value of the relative frequency in the sample, but also on the way the sample has been obtained, and therefore upon the *general method* of sampling employed. For example, the mere fact that there is a high proportion of French-speaking individuals in a sample of 10,000 selected, say, from the residents in a border town in northern Vermont, does not, by itself, provide competent support for the hypothesis that there is a correspondingly high relative frequency of French-speaking residents in the U.S.

Even more dubious, however, is the assumption that if h is the hypothesis that a still unexamined individual has a given property M, the degree of evidential support for h stands in some determinate relation to the relative frequency with which M occurs in a sample. To take an extreme case first, if in a sample consisting of just *one* bird of a given species the bird is found to have a yellow plumage, so that the relative frequency of this property in the sample is 1, no one but a tyro in inductive procedure would assign a high degree of support to that evidence for the hypothesis that the next individual of that species will also have that property. In such a case, barring the use of some well-established theory, we would, I think, be inclined to regard the evidence as insufficient to warrant any conclusion; and if we did assign a degree of support to it, it would, I suspect, be vanishingly small.

But consider a less fanciful example, and suppose we found that in 10,000 observed cases of human births 53 percent are male. Is it plausible to say that the degree of support this evidence gives to the hypothesis that the next human birth will be male is even approximately equal to .53? Since we do not, in actual practice, assign numerical measures to degrees of evidential support, it is not obvious how to answer this question. But it seems to me that here too much would depend on what method of sampling has been employed in obtaining the evidence, on whether we have any reason to believe that the sample is representative of the population, and on the care with which the data have been collected and recorded. The mere fact that the relative frequency of male births in a sample is .53 does not necessarily indicate how cogent is the evidence of that sample for the hypothesis in question. It is certainly not absurd to believe that though the relative frequency of male births in two samples of 10,000 each is the same, the evidential support for the hypothesis given by one sample may be much higher than the support supplied by the other.

Carnap seeks to show, however, that the degree of confirmation for a hypothesis can be construed as the "fair betting quotient" on that hypothesis, and also as an estimate, based on the evidence, of the relative frequency with which a property occurs in the population. He defines a "bet" as a contract between two parties X_1 and X_2, which stipulates that X_1 will confer a benefit u_1 on X_2 if a certain prediction h is fulfilled, while X_2 will confer a benefit u_2 on X_1 if not-h is realized. The ratio

$$q = \frac{u_1}{(u_1 + u_2)}$$

is called the betting quotient; and a betting quotient q is said to be "fair" if it does not favor either party—if, that is, given the evidence e, betting on h with odds q is as good a choice as betting on not-h with the odds $(1\text{-}q)$. Carnap then reasons as follows. Suppose X_1 and X_2 make n similar bets, with the betting quotient q, on each of the n individuals in a class K having the property M. Let h be the hypothesis that a certain individual a in K has M; and assume that the evidence e, available to both parties, asserts the ratio of individuals in K with the property M to be r. Since X_1 will obviously win rn bets with a gain of rnu_2, and will lose $(1\text{-}r)n$ bets with a loss of $(1\text{-}r)nu_1$, his total gain or loss will be

$$n(u_1 + u_2)(r - q)$$

and since he will break even if $q = r$, the betting quotient will be fair when $q = r$. Carnap therefore suggests that the statement "The probability$_1$ (or degree of confirmation) of h with respect to the evidence e has the value q," can be interpreted to say that a bet on h with a betting quotient q, is a fair bet when the bets are placed on the basis of the information contained in e. And he concludes that "If the relative frequency of M in a class to which a belongs is known to be r, then the fair betting quotient for the hypothesis that a is M, and hence the probability$_1$ of this hypothesis, is r."[13]

It is nevertheless not at all clear why the evidence e, which asserts that the relative frequency of M in K is r, should be assigned a probative force of degree r, for the hypothesis that an arbitrary member of K has M. Such an assignment is plausible when K is finite and r is 1 or 0; but for other values of r, it is only on the ground of some obscure continuity assumptions that r can be taken as the measure of evidential support which e gives to h. Moreover, it seems to me that it is only on the basis of a tenuous analogy that Carnap's reasoning in this context can be extended to the interpretation of his probability$_1$ when the hypothesis is a *universal* statement—for in this latter case the bets would have to be placed on possible universes. What does appear plausible is that r is a measure, not of the degree of support given to h by e *simpliciter*, but of the effectiveness in the long run of the *method of placing bets* on evidence such as e. On this interpretation, however, the degree of confirmation which the evidence e lends to h becomes identical with the relative frequency with which a certain method of inference yields true conclusions of a certain form from premises of a certain type. But such an interpretation is not congruous with Carnap's outlook, since the interpretation seriously compromises the purely logical (or analytical) character that statements about degrees of confirmation must have in his view.

Carnap also argues, however, that the degree of confirmation can be construed as the *estimate* of a relative frequency. Thus, suppose that in the above betting situation the evidence e does not contain information about the relative frequency of M in K. He then maintains that "Since the probability$_1$ of h on e is intended to represent a fair betting quotient, it will not seem implausible to require that the probability$_1$ of h on e determines an *estimate of the relative frequency* of M in K."[14] But the estimate based on e of the unknown value of a magnitude u is defined as the mathematical expectation of u—i.e., as the sum of the products formed by multiplying each possible value of u with the probability$_1$ of this value on the evidence e. Since K contains n individuals, there are $n + 1$ possible values for the unknown relative frequency r of M in K. Let r' be the estimate of r. Then the estimated gain or loss g' to the bettor X_1, were he to place n simultaneous bets with X_2, is

$$n(u_1 + u_2)(r' - q)$$

where q is the betting quotient. And the bets will be fair when the estimated gain or loss is 0, so that r' must be equal to q.[15]

But here again the proposed identification of c with an estimate of relative frequency is made on assumptions that cannot be plausibly extended to all the cases to which Carnap's notion of probability$_1$ is intended to apply. One of these assumptions is that, provided the evidence e supplies no information about the individual members of K, it makes no difference which class K is under consideration when placing bets or how many individuals K contains.[16] Another assumption is that *simultaneous* bets are placed on all the members of K, so that e need contain no information about the way in which K is sampled—indeed, Carnap allows for the possibility of forming an estimate of relative frequency even when e is a purely logical truth and no sampling whatever has been made. But when an estimated relative frequency is formed on such assumptions, does the estimate constitute a reasonable measure for the support given by the evidence to the hypothesis that a specified member of the class has a certain property? Suppose, for example, the hypothesis asserts that the next human birth will be a male, and the evidence states that in a certain town with a population of 500 mothers, 300 gave birth to boys. We may of course form an estimate of the relative frequency of male births. I do not think, however, that if we do so we would accept this estimate as to the measure of the support which the evidence gives to the hypothesis. For the degree of support seems to me negligible, while the numerical value of the estimate is conceivably a fairly high ratio. Again, the assumption that simultaneous bets are placed on all members of a class is a bit of fiction that has verisimilitude in certain contexts, but appears to be quite unrealistic for most situations in which we form estimates of relative frequencies and weigh the evidential support for hypotheses. Moreover, Carnap requires his c to be capable of a completely general application, so that the hypothesis h might, in principle, be a state-description (i.e., a complete description of a "possible world"). In such a case, at any rate, placing bets on h would be to bet on something concerning which we could in fact not know whether or not it is realized; and in consequence, the analogy to ordinary betting would be stretched to the breaking point.

Thus far, my comments have been addressed to that part of Carnap's assumption which makes the empirical factors s_M/s one of the end-points of the interval into which the value of $c(h_M, e_M)$ is postulated to fall. What about the remaining part of the assumption, that the other end-point is the logical factor w/k? Carnap maintains that "Other things (including k) being equal, if w/k is greater in one case, the value of c is higher or equal. . . . I believe that the stronger statement with 'higher' instead of 'higher or equal' also holds."[17] He defends this stronger condition by showing that otherwise a logical possibility would be ignored that the evidence e does not exclude.

The central issue which this part of the assumption raises is whether, in the absence of any factual evidence for a hypothesis, it is nevertheless reasonable to assign a degree of support for it on the "evidence" of a purely logical truth. For if $c(h_M, e_M)$ is in general equal to

$$\frac{s_M + \lambda(w/k)}{s + \lambda},$$

then when $s = 0$, c is equal to w/k. The rationale for assigning such a value to c in this case appears to be the following. Since the logical width of M is w, then if an arbitrary individual a is to possess M, a

must fall into one of the *w* narrowest classes specifiable in the language adopted. There are, however, *k* such classes. In the absence of any empirical evidence, is it not therefore plausible to take w/k as the measure of *c* for the hypothesis that *a* is M? Indeed, such reasoning appears to underlie assumptions commonly made in inductive reasoning. Consider, for example, three empirical generalizations having the forms G_1: All A is B; G_2: All AC is B; and G_3: All A is BD, where 'A', 'B', 'C', and 'D' are independent primitive predicates. It is sometimes said that even before empirical evidence is available for any of these generalizations, G_2 is initially "more likely" than G_1, and G_1 is initially "more likely" than G_3. The reason usually given for this claim is that G_1 asserts B to belong to everything that is A, while G_2 asserts B to belong only to the things that are both A and C—so that G_2 asserts less than what is asserted by G_1, and in consequence we are risking error to a lesser degree in accepting G_2 than in adopting G_1. Similarly, G_3 asserts that both B and D belong to everything that is A, while G_1 asserts only that the property B alone belongs to the A's, so that again less is being asserted by G_1 than by G_3. All this is in agreement with Carnap's stipulations, for if we calculate the relative logical widths of the predicates in these generalizations in the manner he describes, G_1 has the relative logical width ¾, G_2 has ⅞, and G_3 has ⅝. For G_1 to be true, ¼ of all the narrowest classes formulable in the language must be empty, for G_2 to be true ⅛ must be empty, and for G_3 to be true ⅜ must be empty. Accordingly, antecedent to any empirical evidence, there appears to be a "better chance" that G_2 is true than that G_1 is, and a "better chance" that G_1 is true than that G_3 is.

Nevertheless, however persuasive these considerations may seem, one must not ignore the fact that in all this we are simply counting logical possibilities. Unless there are some reasons for supposing that some of these possibilities are actually realized, their mere number does not constitute a relevant factor in estimating the evidential support for a hypothesis. For example, if in the case of G_2 there are in fact no individuals which are both A and not-C, the class determined by $A \cdot C$ coincides with the class determined by A alone; and in that eventuality it would be hard to imagine a good reason for regarding G_2 as "more likely" than G_1. But if we know neither that the class of things which are $A \cdot -C$ has any members nor that the class is empty, it is only on the most far-fetched assumption that it is plausible to assign a greater antecedent degree of confirmation to G_2 than to G_1—on the assumption, for example, that our actual universe is a random instance of possible universes, each of which is generated by filling in various proportions the narrowest logically determinable classes formulable in our language.

To make this point for Carnap's actual assumption about w/k as one end-point of the interval into which $c(h_M, e_M)$ must fall, suppose that M is the property of possessing brown eyes, black hair and a light complexion, and that its relative logical width is ¾. Then antecedent to any factual information as to whether M is physically realizable, or if so realizable whether there are any individuals with M, $c(h_M, e_M)$ must equal ¾. How can this value be justified? It would be justified if we had some reason for supposing that, in Charles Peirce's words, universes are as plentiful as blackberries, and that were we to bet on the occurrence of an individual with M in each of the universes (though without possessing any factual information about any of them) we would win in approximately ¾ of the time. Since, however, such a supposition is at best fanciful, it cannot really serve to justify the value assigned to c.

But let us modify the example somewhat, and assume that in a sample of 100 individuals 80 are found to have M. Then $c(h_M, e_M)$ must, on Carnap's stipulations, be not less than ¾, irrespective of the way in which the sample has been obtained—even if a sampling procedure is used which does not in general yield representative samples, so that the ratio 4/5 of M in the actual sample might happen to be far greater than the ratio of M in the population. It does not seem to me, however, that under the circumstances a value for c equal to or greater than ¾ is either a plausible or a usable measure for the support which the evidence lends to the hypothesis. And I do not believe, therefore, that a good case has been made out for stipulating that in general the value of $c(h_M, e_M)$ must fall into the interval determined by the empirical and logical factors.

8

But even if Carnap's stipulations for constructing a continuum of inductive methods are granted, the problem still remains of choosing one of the methods as in some sense "the best." Carnap himself rejects the method specified by taking the parameter $\lambda = \infty$ on the ground that since this method gives an "infinite weight" to the logical factor, no amount of factual knowledge could alter the a priori degree of confirmation for a hypothesis, a result which is clearly incompatible with scientific practice. Carnap nevertheless maintains that the choice of a method is not a "theoretical" question, because the selection of a method involves a practical decision into whose determination a variety of nonfactual considerations will enter. Thus, he maintains,

> A possible answer to a theoretical question is an assertion; as such it can be judged as true or false, and, if it is true, demands the assent of all. Here, however, the answer consists in a practical decision to be made by X. A decision cannot be judged as true or false but only as more or less adequate, that is, suitable for given purposes. However, the adequacy of the choice depends, of course, on many theoretical results concerning the properties of the various inductive methods; and therefore the theoretical results may influence the decision. Nevertheless, the decision itself still remains a practical matter, a matter of X making up his mind, like choosing an instrument for a certain kind of work.[18]

It will be clear, however, that the choice of a method is not a theoretical question, only as long as the objectives which an inductive logic is to achieve are not specified. Once these objectives are made explicit, there surely are factual grounds for preferring one method (or group of methods) over others. Indeed, Carnap himself recognizes this implicitly when he rejects the method $\lambda = \infty$, and explicitly when he declares:

> Suppose that X has chosen a certain inductive method, and used it during a certain period for the inductive problems which occurred. If, in the view of the services this method has given him, he is not satisfied with it, he may at any time abandon it and adopt another method which seems to him preferable. . . . How can X go over from one inductive method to another? It is not easy to change a belief at will; good theoretical reasons are required. It is psychologically difficult to change a faith supported by strong emotional factors (e.g., a religious or political creed). The adoption of an inductive method is neither an expression of belief nor an act of faith, though either or both may come in as motivating factors. An inductive method is rather an instrument for the task of constructing a picture of the world on the basis of observational data and especially of forming expectations of future events as a guidance for practical conduct. X may change this instrument just as he changes a saw or an automobile, and for similar reasons. . . . After working with a particular inductive method for a time, he may not be quite satisfied and therefore look around for another

> method. He will take into consideration the performance of a method, that is, the values it supplies and their relation to later empirical results, e.g., the truth-frequency of predictions and the error of estimates; further, the economy in use, measured by the simplicity of the calculation required; maybe also aesthetic features, like the logical elegance of the definitions and rules involved. The λ-system makes it easy to look for another inductive method because it offers an inexhaustible stock of ready-made methods systematically ordered on a scale. . . . Here, as anywhere else, life is a process of never-ending adjustment; there are no absolutes, neither absolutely certain knowledge about the world nor absolutely perfect methods of working in the world.[19]

The crucial point to be noted in these comments is that if we wish to choose an inductive method which will be consonant with the aims of empirical science, we must take into consideration "the performance of a method, that is, the values it supplies and their relation to later empirical results, e.g., the truth-frequency of predictions and the error of estimates." For the question now arises in what way "later empirical results" and the truth-frequency of predictions are relevant for judging a method's "performance." On the basis of what has been said thus far, however, it is difficult to see that these things *are* relevant in the slightest degree. For in Carnap's system, if $c(h, e) = p$ is true, it is an analytic truth, invulnerable to the facts of experience; and even though h should turn out to be false while p is high, nothing in this circumstance can affect the validity of the statement that p is the value of $c(h, e)$. Evidently, then, some further postulates are required—for example, a postulate such as that if h is a singular prediction and $c(h, e) = p$, then in a class of similar predictions made on the basis of analogous evidence the predictions turn out to be true with a relative frequency close to p. If such additional postulates are introduced, however, an inductive logic is in effect converted into an *empirical theory* about the success-ratios of predictions concerning the course of events. Though the logical structure of the predictive mechanism can be analyzed without reference to such events, the system which includes these additional postulates is no longer simply a branch of pure mathematics.

Carnap appears to recognize this in so many words. In his preliminary comments on the choice of an inductive method, he imagines a system of repeated simultaneous bets by two persons on the hypothesis that an individual not in the sample has some property M, where each person bases his wager on the evidence supplied by a sample of fixed size s but each uses a different c-function. (It is assumed that the number of successful bets each person makes can be ascertained.) Carnap supposes, furthermore, that such bets are made for all other properties as well, and that samples of all other (finite) sizes are employed as evidence. He then suggests that the overall balance of gains or losses for the total system of bets based on one c-function could be taken as the measure of the relative success of that inductive method, as compared with the analogous measure of success when a different c-function is used for placing such a system of bets. This is a daring though not immediately promising suggestion, for it proposes a measure for the merit of an inductive method that could be evaluated only if we had complete knowledge of the contents of our universe. Nevertheless, the suggestion does indicate that in Carnap's view, *theoretical* and not merely practical questions do enter into the choice of satisfactory inductive method. The problem remains, however, whether Carnap can indicate an effective basis for choosing between the different alternatives in his continuum of inductive methods.

In point of fact, Carnap works out in a technically impressive way the central idea contained in these preliminary suggestions. On the assumption that the actual universe has a certain given constitution (i.e., on the assumption that a given state-description U is the true one), he develops a formula for the optimum value of the parameter λ. This formula is obtained as follows. Suppose that the k narrowest classes distinguishable in the language adopted are determined by the properties Q_i ($i = 1, 2, \ldots, k$), and that the relative frequency with which Q_i occurs in the universe is r_i. Suppose, further, that estimates are made of the r_i on the basis of all possible samples of fixed size s drawn from the universe, where the estimating function itself is given in terms of some definition of c (and therefore in terms of some value of the parameter λ). Now it is customary to regard an estimating function as the more successful, the smaller is the mean-square of the deviations of the estimates from the true value of the magnitude being estimated. It turns out that this mean-square for all the r_i's, where the estimates are based on samples of fixed sixe s, is equal to

$$\frac{s - (\lambda^2/k) + (\lambda^2 - s)\,\Sigma\, r_i^2}{k(s + \lambda)^2},$$

which can therefore be taken as the measure of the effectiveness of the inductive method λ. In this formula, the sum Σr_i^2 is taken by Carnap to measure the degree of "order" or "uniformity" in the universe. This sum has the maximum value 1, when all the individuals in the universe fall into one of the narrowest classes (i.e., when they all possess just one of the Q-properties); it has the minimum value $1/k$, when the individuals are distributed equally among those classes. It immediately follows that when Σr_i^2 has neither the maximum nor the minimum value, the optimum inductive method for a given state-description U is given by

$$\lambda\Delta = \frac{1 - \Sigma r_i^2}{\Sigma r_i^2 - 1/k}.$$

But this result is of no value for actually choosing an inductive method, since even if we can persuade ourselves that we know the value of k (which is a function of the number of primitive predicates required for describing the universe), we certainly do not know the value of Σr_i^2 for our actual universe. As Carnap notes, "The practical knowledge situation for any human being at any time is such that he knows only a relatively small part of the universe, never the whole of it; it is just this fact that makes the use of inductive methods necessary."[20] Nevertheless, although Carnap leaves unanswered the question how then we are to choose the optimum inductive method, he believes it is possible to fix a lower bound for $\lambda\Delta$, and so eliminate certain inductive methods as not optimal.

He reasons as follows. Let U_T be the unknown true state-description, and $r_{iT} = N_{iT}/N$ be the unknown relative frequencies with which the k Q-properties occur in it. Suppose a sample of size s is drawn, and that it is nonhomogeneous—i.e., at least two distinct Q-properties occur in it. Then the universe itself cannot be homogeneous. If the sample contains s_i individuals which are Q_i, clearly $r_{iT} \geq s_i/N$. Suppose now that the property Q_m occurs most frequently in the sample, so that s_m/s is greater than any other relative frequency in the sample. Now construct a state-description U which will agree with the information in the sample, but which assigns to all individuals not included in the sample the property Q_m. Accordingly, if N_i' is the number of individuals in U with the property Q_i, the relative frequency of Q_i in U is given by $r_i = N_i'/N$. For $i \neq m$, $N_i' = s_i$; while for $i = m$, $N_m' = N - s + s_m$. It follows that $\Sigma r_i^2 < 1$. But the unknown true U_T cannot be nearer to homogeneity than U; hence

$$\Sigma r_{iT}^2 \leq \Sigma r_i^2 < 1,$$

and Σr_i^2 is an upper bound for Σr_{iT}^2. Carnap therefore declares that Σr_i^2 is "a *known* upper bound" for the unknown Σr_{iT}^2, so that the unknown optimum method λ_T^Δ must be not less than

$$(1 - \Sigma r_i^2)/(\Sigma r_i^2 - 1/k)$$

and must therefore be greater than 0.

But it is clear that the value of Σr_i^2 can be obtained only if N is known, and also that the value of $\Sigma r_i^2 - 1/k$ is known only if both N and k are known. In point of fact, however, neither N nor k is known, and to assume that they are is to assume something about "the practical knowledge situation for any human being at any time" which Carnap himself would doubtless admit is contrary to the facts. Moreover, if N is very large when compared with s (the size of the sample),

$$r_m = N_m/N = 1 - (s - s_m)N,$$

which is close to one; in consequence, Σr_i^2 itself will approach 1 with increasing N. Accordingly, for very large N (and there is good reason to believe that N is quite large when compared with the size of samples drawn from our actual universe), $1 - \Sigma r_i^2$ will be close to 0. It follows that even if the optimum method λ_T^Δ for the actual universe is admitted to have a lower bound greater than zero, this lower bound is for all practical purposes indistinguishable from 0. It does not seem, therefore, that Carnap has provided any usable clues for choosing between the inductive methods in his continuum. Indeed, so many fundamental questions are left unanswered in his system, so many appear impervious to a reasonable resolution, that I am forced to regard with grave scepticism its significance as a potential clarification of inductive practice.

9

Carnap does not dismiss the problem of "justifying" induction as a spurious one. Though he touches on the question only briefly, the reasons he gives for requiring such a justification, as well as the way he thinks it can be supplied, throw further light on his general philosophy of inductive inference. I wish now to comment on these matters.

Carnap notes that we cannot know with certainty whether a prediction (e.g., it will rain tomorrow), based on given evidence, is true before the event. Lacking such certainty, we must, according to him, adopt an appropriate course of action in the light of statements like "With respect to the available evidence, the probability, that it will rain tomorrow is high." On Carnap's view, however, the latter statement expresses a truth of logic; and if this is so, the obvious question is why we should base our actions on it. Now we may have some ground for believing that though this single prediction may be falsified by the events, similar predictions will in the long run have a high success-ratio. Accordingly, the issue reduces itself to finding good reasons for a thesis such as the following: "If X makes a sufficiently long series of bets, where the betting quotient is never higher than the probability$_1$ for the prediction in question, then the total balance for X will not be a loss"—and more generally, if X makes his decisions by taking into account the values of probability$_1$, "he will be successful in the long run."[21]

As Carnap is careful to point out, however, this thesis is not a truth of logic, and is warranted only "if the world as a whole had a certain character of uniformity to the effect, roughly speaking, that a kind of events which have occurred in the past very frequently under certain conditions will under the same conditions occur very frequently in the future." He thus accepts the orthodox view that a "presupposition" of induction (i.e., of basing our actions on the value of probability$_1$) is the familiar doctrine of the uniformity of nature—a doctrine which he renders as "The degree of uniformity of the world is high."[22] But since this doctrine is in turn a factual hypothesis, and ostensibly underlies all inductive inference, many thinkers have argued that the doctrine cannot be established by inductive reasoning, on pain of circularity or an infinite regress; and they have felt compelled to advocate either scepticism or the abandonment of empiricism.

Carnap, on the other hand, does not believe this argument to be sound or the alternatives it presents to be exhaustive. For he maintains that it is not essential for the justification of induction that we know *with certainty the truth of the doctrine of the uniformity of nature. All that is needed, according to him, is that we establish the claim that "On the basis of the available evidence it is very probable* [in the sense of probability$_1$] that the degree of uniformity of the world is high."[23] This claim, then, is the presupposition sufficient for establishing the validity of induction. But this presupposition is no longer a factual hypothesis; and if it is true then, like all probability_1 statements on Carnap's view, it is a truth certifiable by logic alone—though the proof of this logical truth is reserved by Carnap for a later publication. He therefore concludes that we cannot obtain a reasoned assurance that we will be successful in the long run by using inductive methods; but we can have demonstrated assurance that it is *probable* that we will be successful, since the assertion that the success is probable is analytically true.

Is it not a puzzle, however, how an allegedly analytic truth, which asserts nothing about the constitution of the actual world, can serve as the basis for practical decisions about the actual course of events? Carnap's reply to this natural query is that the analytic truth in question simply *makes explicit* the inductive logical relations between the available evidence and the hypothesis of nature's uniformity: the analytic truth merely shows the high probability of the hypothesis relative to the evidence, and thereby makes evident that it is reasonable for us to act on high probabilities. And Carnap concludes:

> A practical decision is reasonable if it is made according to the probabilities with respect to the available evidence, even if it turns out to be not successful. . . . It is reasonable for X to take the general decision of determining all his specific decisions with the help of the inductive method, because the uniformity of the world is probable and therefore his success in the long run is probable on the basis of his evidence, even though he may find at the end of his life that he actually was not successful and that his competitor who made his decisions in accordance not with probabilities but with arbitrary whims was actually successful.[24]

I now turn to my comments. Carnap adopts what is in effect a species of deductive justification of induction, one in which all the premises and therefore the conclusion are allegedly analytic truths. I shall not question his claims that within his system it is an analytic truth that the high degree of uniformity of the world is probable on the evidence; for in the absence of a published proof of the claim it would be pointless to do so.[25] But I do question whether his proposed justification can dispense with some factual assumptions, however disguised these assumptions may be by having them appear as conventions built into the structure of the language which is adopted. Thus, I have already noted the similarity between the Baconian doctrine of forms and Keynes's principle of limited independent variety, on the one hand, and Carnap's assumption that a finite number of primitive predicates is sufficient to describe the world completely. This assumption, like those of Bacon and Keynes, is surely a factual one; and though there may be good evidence for it, the degree of support which the evidence gives to it cannot be estimated, on pain of circularity, in terms of the inductive logic based on that assumption. If this point is well-taken, however, Carnap is faced with the problem, traditional to all attempts at a deductive justification of induction, of validating what is ostensibly the supreme factual major premise in all inductive reasoning. But as far as I am aware, none of the tools thus far included in his armory of logical devices is adequate to this task.

It is pertinent to ask, therefore, whether in fact a supreme factual major premise is needed for justifying induction—or alternatively, whether a deductive justification of induction in a wholesale fashion is a reasonable undertaking. As Carnap tacitly recognizes, no major premise of the required generality can be specified which would permit the deduction of an inductive conclusion from such a premise, even when the latter is combined with instantial evidence. The most one can hope to achieve in this connection is to deduce from the combined premises, not the inductive conclusion itself, but the statement that the inductive conclusion is probable to some degree on the evidence. Such a deduction will be effected, however, by way of the application of some *rule of inductive inference*, in accordance to which the deduction must be made; and presumably different rules will be required for different types of conclusions and evidential premises. The validity of inductive arguments is then reduced to the question as to the warrant for assigning, in accordance with such rules, stated degrees of probability to inductive conclusions based on given premises. But in any event, since a variety of rules are needed, the inductive conclusions may just as well be inferred from the specific, evidential premises alone, in accordance with the rules referring to such premises exclusively, instead of being inferred from premises that include an otiose additional "major" assumption. For these reasons a wholesale deductive justification of induction seems entirely gratuitous.

But to return to Carnap. According to him, every statement which assigns a degree of confirmation to a hypothesis on given evidence is analytic; and the analytic character of such statements is indeed a consequent of the fact that the assignment of degrees of confirmation is made in accordance with stated definitions and rules. Once such rules are granted, the claim that statements about degrees of evidential weight are analytic seems to me to hold, not only for Carnap's precisely articulated system, but also for the looser evaluations of evidence we actually make in science and ordinary affairs of life. If we judge, for example, that the evidence consisting of a rolling gait, a bronzed complexion, and calloused hands confirms to a high degree the hypothesis that a certain person is a sailor, the judgment involves the application to the case at hand of some conception, however vague it may be, of what constitutes good evidence. Should two individuals disagree on the cogency of this evidence, the disagreement seems to follow either from the fact that one of the individuals possesses unmentioned items of evidence which the other lacks, or from the fact that the individuals are employing different standards of what constitutes good evidence. In the former case, the dispute can be settled by making explicit the full range of evidence upon which the differing judgments are made. In the latter case, the dispute can be settled only by making explicit the different standards employed, and so recognizing that evaluations of the same evidence in terms of differing standards cannot but be different. If, however, the available evidence and the standards of evaluation are common to the individuals, then, except for possible blunders in applying the standards, their judgments on the merit of the evidence for the given hypothesis will coincide—and no appeal to empirical data other than that which is mentioned in the evidence seems relevant in making the judgments. Under these conditions, accordingly, statements about degrees of evidential support are undoubtedly analytic.

The question remains, however, whether the standards employed in evaluating evidence—i.e., the rules used in assigning degrees of confirmation to hypotheses relative to stated evidence—are not themselves based on factual assumptions, and do not themselves require to be defended by some appeal to empirical data. The answer, in my opinion, is strongly affirmative, at any rate for the rules explicitly or tacitly employed in weighing evidence in actual inductive practice. For example, many inductive inferences proceed in accordance with the rule (induction by simple enumeration) that a conclusion about the composition of a class of elements may be asserted on the basis of the composition of samples drawn in a certain manner from that class. This rule cannot be defended on a priori grounds. It can be justified only on the ground that the ratio of successful inferences drawn in the past in accordance with the rule is actually high. More generally, while I do not believe that induction can be justified in a wholesale and once-for-all fashion, I think that individual inductive policies can be justified—and indeed justified without vicious circularity—in terms of the *de facto* success-ratios that are associated with the inductive rules underlying those policies.

Where does Carnap stand on this issue? I do not find his published writings entirely clear on the point. As some of the quotations cited above indicate, he does say that an inductive method is sound if, in making our decisions in the light of the degrees of confirmation the method specifies, we are "successful in the long run." Such remarks certainly suggest that Carnap subscribes to the conception of the justification of induction I have just outlined. But on the other hand, he also maintains that a man is acting "rationally" in basing his decisions on the degrees of confirmation prescribed by an inductive method, "even though he may find at the end of his life that he was not successful." If it is safe to assume that Carnap intends this dictum to cover not only the arbitrary period of one man's life, but also the longer period of the life of the human race (and perhaps an even more extended stretch of time), he is then committed to what is surely a curious notion of "rationality"—and he must then be understood as rejecting an interpretation of his views such as the one just mentioned. But in any event, I do not believe that a man would ordinarily be judged to be acting in a rational manner, provided only he conducts his affairs consistently in accordance with some fixed set of rules and irrespective of the consequences such action brings forth. Consistency is undoubtedly an ingredient in our conception of rationality; but while it may be a necessary requirement for rational behavior, it is hardly a sufficient one. If rationality is conceived in the manner in which Carnap appears to conceive it, probability ought surely not to be taken as the guide to life.

This is, I must admit, a most ungracious essay. For if the major criticisms advanced in it hold water, it shows that despite the remarkable constructive power and ingenuity Carnap has brought to the reconstruction of inductive logic, he has not resolved the outstanding issues in the philosophy of induction, and his general approach to the problems is not a promising one. My excuse for writing this essay is not only that a forthright critique of his work in this domain is itself a testimony to my profound admiration for the intellectual power his work exhibits. My excuse is, further, that Carnap may perhaps find useful for the completion of his magnificent system a statement of some of the difficulties which a sympathetic reader encounters in the fragments already published, and that he may perhaps be stimulated to show how those difficulties can be outflanked.

8/ "Impossible Numbers": A Chapter in the History of Modern Logic

The distinctions, currently recognized, between "pure" and "applied" mathematics, and the identification of the former with logic, is frequently attributed to George Boole, when not regarded as an achievement of contemporary logicians. Very little notice is taken of the fact that Boole was a mathematician of note who shared the philosophy of mathematics of many English as well as continental thinkers of his day.[1] A clear indication that his work in symbolic logic was simply the application to new fields of the dominant conceptions of contemporary mathematicians, is to be found in the Introduction to Boole's *Mathematical Analysis of Logic.*[2]

> They who are acquainted with the present state of the theory of Symbolical Algebra are aware, that the validity of the processes of analysis does not depend upon the interpretation of the symbols which are employed, but solely upon the laws of their combination. Every system of interpretation which does not affect the truth of the relations supposed, is equally admissible, and it is thus that the same process may, under one scheme of interpretation, represent the solution of a question on the properties of numbers, under another, that of a geometrical problem, and under a third, that of a problem of dynamics or optics. This principle is indeed of fundamental importance; and it may with safety be affirmed, that the recent advances of pure analysis have been much assisted by the influence which it has extended in directing the current of investigation.
>
> But the full recognition of the consequences of this important doctrine has been, in some measure, retarded by accidental circumstances. It has happened in every known form of analysis, that the elements to be determined have been conceived as measurable by comparison with some fixed standard. The predominant idea has been that of magnitude or more

> strictly of numerical ratio. The expression of magnitude, or of operations upon magnitude, has been the express object for which the symbols of Analysis have been invented, and for which their laws have been investigated. Thus the abstractions of the modern Analysis, no less than the extensive diagrams of the ancient Geometry have encouraged the notion, that Mathematics are essentially, as well as actually, the Science of Magnitude.

Boole's application of mathematical methods to logic must be viewed, therefore, as the working out of the idea that the proper and exclusive subject matter of mathematics is *not quantity*. But the insights into the nature of mathematics which Boole here records did not originate with him, however much he may have contributed to their extension. These were won only after more than a hundred and fifty years of persistent effort on the part of mathematicians to see clearly what was the nature of their science, but which did not find tolerably complete expression before Peacock, Gregory, DeMorgan, Hamilton, and Boole resolved to establish algebra on a firm foundation. The story, or at least the partial story, of the clarification of the understanding concerning what mathematics is, is the theme of this paper.

Knowledge of the historical development of certain ideas, however, is not the only thing which can be won from the study of that development. Not only is the point of contact between recent tendencies in logical theory and a long tradition in mathematics thrown into relief; but I think that the current distinction between "pure" and "applied" mathematics, since it receives ample illustration in the struggles of mathematicians with the principles of their science, is itself clarified. The attempted "identification" of pure mathematics and logic, and the "deduction" of the former from the latter, often appears not as the intelligible process of *exhibiting* the most general principles of a determinate subject matter, but as the unintelligible attempt to *derive* a determinate subject matter from principles so general that they are not exhaustive of any special subject matter at all. For unfortunately, the term "pure mathematics" is not always used unambiguously, and is sometimes even identified with what must be regarded, with good reason, as *applied* mathematics, a branch of natural science: e.g., when by numbers, lines, etc. and their properties, reference is made to the structure of the activities involved in counting, measuring, etc. On the other hand, the concern of many mathematical logicians with the structure of "symbolical" systems, appears to some like a myopic preoccupation with "meaningless marks," so that the bearing of these algebras on a philosophy of mathematics is misunderstood. No doubt a clear apprehension of these matters may be won without recourse to the history of the issues involved. Nevertheless, by indicating the historical setting for current logical theories, even those in fundamental disagreement with them may acquire a richer appreciation of the motivation for their occurrence, and so of the difficulties which these theories are designed to meet.

1

Prior to the first quarter of the nineteenth century, mathematics was universally defined as the science of quantity. It was under the domination of this idea that Euler wrote his influent algebra.

> Everything will be said to be a magnitude [he stated], which is capable of increase or diminution, or to which something may be added or subtracted. Thus a sum on money is a magnitude. . . . There are many different kinds of magnitude, which cannot be all enumerated; on this account there are different branches of mathematics each of which studies a different kind of magnitude, since mathematics is nothing more than the *science of magnitudes*.[3]

Accordingly, "number" is conceived as being, in first and last instance, an answer to the questions "How many?" and, in cases of extensive measure, "How much?"

For a long time this view represented accurately the subject matter of mathematics. Even so, minor crises occurred whenever the range of applicability of the arithmetical operations was extended. For example, multiplication was at first rightly viewed as a repeated addition of concrete units, and since it was defined only for multiplication by abstract integers, it was a process which always *increased* a magnitude. But it soon became imperative to conceive a process of "multiplication" applicable in some sense to ratios of intergers (designated by fractions) as well as to integers. There is a dialogue between master and scholar in the sixteenth-century arithmetic of Robert Recorde, in which the pupil struggles with the difficulty that in some cases multiplication should make a thing *less*. The master finally explains:

> If I multiply by more than one, the thing is increased; if I take it but once, it is not changed; and if I take it less than once, it cannot be so

> much as it was before. The seeing that a fraction is less than one, if I multiply by a fraction it follows that I do take it less than once.

To which the scholar replies with great courtesy, "Sir, I do thank you much for this reason; and I trust that I perceive the thing."[4] But his politeness is more conspicuous than his understanding, and he is quite unaware that in place of the familiar idea of multiplication as repeated addition, a *different*, and if one likes a misleading expression, *a more general* notion had been substituted.

Now while ratios of integers and operations with them in accordance with the same formal rules as with integers, had been accepted as "natural" extensions of the original concept of number, "negative" and "imaginary numbers" were not welcomed so readily. After a general method was discovered for solving algebraic equations of the first degree, it was found that in some cases a peculiar kind of root, namely, a "negative number," was obtained. Such roots clearly did not correspond to anything that had heretofore been recognized as quantity; hence what meaning to assign to these "negative quantities" became a serious problem. But the question of "negative quantities" was only child's play when compared with the problem of "imaginary" or "impossible numbers" which involved the square root of negative unity. For expressions involving these turned up as inevitably as "negative numbers" did, when otherwise reliable general methods were used to solve certain quadratic equations.

The problem of these "impossible numbers" is an instance of one of the oft-recurring "crises" in the history of science. Its resolution led to a radical overhauling of traditional notions concerning mathematics, to an enlarged view of formal logic, and to their identification in recent days. Meanwhile, however, mathematicians were in a dilemma. If mathematics was the science of quantity, then these "impossible numbers" must be quantities; but "negative" and "imaginary quantities" were simply nonsense, incompatible with all previous definitions of quantity. On the other hand, these "impossible numbers" were the logical consequences of using certain general and powerful methods for solving equations; and the queer numbers could not be banished without calling to an abrupt halt the gradual perfecting of mathematical techniques. "Impossible numbers," absurd as they were, were undeniably useful, as Descartes recognized in making the rules of algebra general, and in making evident important analogies between different problems. Which was to be sacrificed—the logical conscience of mathematicians or the development of the science of mathematics?

Most professional mathematicians, like Maclaurin and Euler, did not hesitate to accept the conveniences which the queer numbers offered at no matter what logical price; although they did try to justify, very lamely indeed, "negative quantities" by illustrating the difference between these and "positive quantities" by the "analogous" difference between owing and owning money, etc.[5] But for a long time no one could defend the "imaginary numbers" with any plausibility, except on the logically inadequate ground of their mathematical usefulness.[6] At best, the occurrence of an "imaginary root" to an equation was taken to indicate that the problem proposed was impossible of solution because it involved a contradiction in terms. At worst, "imaginary quantities" were frankly accepted as a "mystery," and used to support the mysteries of orthodox Christianity.[7]

But if "number" was nothing else than the answer to "how many" or "how much," "negative" and "imaginary numbers" had no place in mathematics. And those who refused to admit them unquestionably had logic on their side, even though in their fierce logical fanaticism they were less wise than those who did. There were many who could not compromise with their logical consciences. Robert Simson, an excellent mathematician and scholar, was one such.[8] And lesser men, like Baron Maseras[9] or William Frend, future father-in-law of DeMorgan and learned anti-Newtonian to boot, stated their views in no mincing language.

Frend's objections are still instructive as well as amusing.

> The ideas of number are the clearest and most distinct in the human mind; the acts of the mind upon them are equally simple and clear. There cannot be confusion in them. . . . But numbers are divided into two sorts, positive and negative; and an attempt is made to explain the nature of negative number, by allusion to book-debts and other arts. Now, when a person cannot explain the principles of a science without a reference to metaphor, the probability is, that he has never thought accurately upon the subject. A number may be greater or less than another; it may be added to, taken from, multiplied into, and divided by another number; but in other respects it is intractable: though the whole world should be destroyed, one will be one, a three will be a three; and no art whatever can change their nature. You may put a mark before one, which it will obey: it submits to be taken away from another number greater than itself, but to attempt to take it away from a number less than itself is ridiculous. Yet this is attempted

> by algebraists, who talk of a number less than nothing, of multiplying a negative number into a negative number and thus producing a positive number, of a number being imaginary. Hence they talk of two roots to every equation of the second degree . . . they talk of solving an equation which requires two impossible roots to make it soluble: they can find out some impossible numbers, which, being multiplied together produce unity. This is all jargon, at which common sense recoils.[10]

Frend rejected, therefore, the principle that every equation has as many roots at its degree, and wrote a treatise which consistently developed the position that algebra was nothing else than a symbolical treatment of *arithmetic,* that is, of *quantities* in the strictest sense. No "negative" or "imaginary numbers" found a place in it, and Frend described those who permit them as

> grave men, imitating the philosophers of a well-known region who were extracting sun-beams from cucumbers, wasting the midnight oil just as profitably in settling the rights and privileges of impossible quantities. The fruits of their lucubrations I sucked in with the first milk of alma mater; and, if the good old lady had not driven me from my books and from studies of a much superior nature, it is not improbable that I might still have looked on these subjects through her spectacles. . . . Being no longer of an age nor of a disposition to take for granted an old wives' tale, I started at the postulate to make a load of corn out of a bushel.[11]

Without question Frend and those like him were in the right in rejecting "impossible numbers," *if* mathematics was the science of quantity. And there is no doubt that criticisms like those of Frend were instrumental in bringing about a theory of "impossible numbers" which not only recognized their usefulness, but supplied for them a logically impeccable foundation. Nonetheless, mathematicians who refused to banish them because such foundations were lacking, were not fools either, as subsequent events showed. For in developing a *calculus* of "impossible numbers," those mathematicians paved the way for the rapid advance of algebra when once these numbers, and the operations upon them, were shown to be as "intelligible" as was arithmetic. Certainly the history of mathematics confirms Mach's penetrating comment, that the "student of mathematics often finds it hard to throw off the uncomfortable feeling that his science, in the person of his pencil, surpasses him in intelligence."

We now turn to the tentative gropings of mathematicians toward the insight that their science does not deal *exclusively* with quantities, and even does not deal with quantities *at all.* We may distinguish two stages in the growth of this view. The first consists in repeated and finally successful attempts to find a geometrical *interpretation* (or "meaning") for "imaginary" as well as "negative numbers." The second consists in the frank recognition that mathematics as a "pure" science is not concerned with any special interpretations of its symbols, but has as its subject matter the structure of systems of uninterpreted signs which are capable of any number of different interpretations.

2

Interpretations for "negative numbers" which were intuitively satisfactory were long familiar, when John Wallis tried to give an equally satisfactory meaning to "imaginary" ones.[12] He made clear, however, that an imaginary root, "as the first and strict notion of what is proposed," denotes nothing more than that a problem is incapable of solution. "For it is not possible, that any Number (Negative or Affirmative) Multiplied into itself, can produce (for instance) −4.' But he hastened to add that

> it is also Impossible, that any Quantity can be *Negative.* Since that is not possible that any *magnitude* can be *Less than Nothing,* or any *Number Fewer than None.*

Nevertheless, Wallis continued, "negative quantities" are useful. As a "bare Algebraick Notation," it is a quantity less than nothing, and so absurd; but as a *sign* having a "physical Application, it denotes as Real a Quantity as if the sign were +." His interpretation of a "negative number" is the familiar one of a length with direction opposite to that of a length symbolized by a "positive number." Similarly, while there cannot "in strictness of speech" be a root of a negative square, such an expression may be interpreted, he argued, no longer, indeed, as a directed line comparable with the two directions of a straight line; but as a segment whose *length* is the mean *proportional* between lengths having opposite directions, and whose direction lies in between these directions, and at some angle or other to the straight line.

Wallis was obviously on the track of a very important idea, namely, that a "bare Algebraick Notation," whether it involved "positive," "negative," or "imaginary" expressions, could be made perfectly intelligible, provided they were not regarded as standing "intrinsically" or "naturally" for quantities. If he was unable to apprehend this fully, it was because he believed that "in strictness of speech" algebraic signs could not refer to more complex structures of relations than hitherto recognized. As it is, he had nothing to say concerning the interpretation of the signs of addition and multiplication of imaginaries. So long as a special interpretation of an algebra is taken as the "natural" or exclusive one, certain combinations of its signs must also be taken as "intrinsically" and absolutely impossible expressions.

This provincialism with respect to the possible interpretation of a calculus still characterized Playfair, when a hundred years later he reviewed the status of the "imaginary numbers." He contrasted the unsatisfactory state of the theory of algebra with the perfection of geometry, and attributed the difference to the fact that the languages employed in these disciplines *symbolized in different ways*.

> The propositions of geometry have never given rise to controversy, nor needed the support of metaphysical discussions. In algebra, on the other hand, the doctrine of negative quantity and its consequences have often perplexed the analyst, and involved him in the most intricate disputations. The cause of this diversity must no doubt be sought for in the different modes which they employ to express our ideas. In geometry every magnitude is represented by one of the same kind; lines are represented by lines, and angles by an angle, the genus is always signified by the individual, and a general idea by one of the particulars which fall under it. By this means all contradiction is avoided, and the geometry is never permitted to reason about the relations of things which do not exist, or cannot be exhibited. In algebra, again, every magnitude being denoted by an artificial symbol, to which it has no resemblance, is liable, on some occasions, to be neglected, while the symbol may become the sole object of attention. It is not perhaps observed where the connection between them ceases to exist, and the analyst continues to reason about the characters after nothing is left which they can possibly express: if then, in the end, the conclusions which held only of the characters be transferred to the quantities themselves, obscurity and paradox must of necessity enduse.[13]

Playfair here almost guessed the secret of the nature of mathematics. But in spite of his acute remarks on signs, his ideas were essentially naive; he did not see that geometric diagrams did no more "naturally" represent configurations in space, than did algebraic diagrams (i.e., signs) "naturally" and necessarily represent quantities.

Hence, "imaginary numbers" were, for him, "paradoxical," even though they can be the subjects of what seemed to be arithmetical operations. But then how explain that such operations upon "imaginary quantities" lead us to important theorems in both algebra and geometry—theorems which can be demonstrated by other, unexceptionable methods?

> Here then is a paradox which remains to be explained. If the operations of this imaginary arithmetic are unintelligible, why are they not also useless? Is investigation an art so mechanical, that it may be conducted by certain manual operations?[14] Or is truth so easily discovered, that intelligence is not necessary to give success to our researches?[15]

Playfair's answer was at best a lame one. Upon examining several instances where "unintelligible" operations lead to significant conclusions, he discovered that the usefulness of these operations depends upon there being certain analogies in form between expressions that involve the imaginaries and other expressions that do not. Thus, for example, he found that the *sine* and the *cosine* of an angle can be stated in terms of the exponential function e with "imaginary" exponents, and hence that many properties of the *circle* can be formulated in terms of "imaginary numbers"; but he also found that when in such expressions $\sqrt{1}$ is substituted for $\sqrt{-1}$, true theorems concerning the hyperbola are obtained. Accordingly, the value of "imaginary numbers" lies in their *suggesting* to us, but not *demonstrating*, analogous theorems for other geometric figures.[16] His final word is that

> the deductions into which they enter are thus reduced to an argument from analogy . . . . . The arithmetic of impossible quantities will always remain a useful instrument in the discovery of truth, and may be of service when a more rigid analysis can hardly be supported.[17]

Thirty years later, in an unsigned review of an important paper of Buée (in which a successful geometric interpretation was given to the arithmetic of imaginaries), Playfair was still priding himself on his refusal to depart from the traditional view of mathematics as the science of quantity.

> The essential character of imaginary expression [he said], is to denote impossibility; and nothing can deprive them of this signification. Nothing like a geometrical construction can be applied to them; they are indications of the impossibility of any such construction, or of anything that can be

exhibited to the senses. . . . There have been more than one attempt to treat imaginary expressions as denoting things really existing, or as certain geometrical magnitudes which it is impossible to assign. The paper before us is one of these attempts: and the author, though an ingenious man, has been betrayed into this inconsistency by a kind of metaphysical reasoning.[18]

And yet, characteristically and amazingly enough, Playfair managed to catch more than a glimmer of the truth which was not to shine brightly for some years:

> The imaginary symbols, considering them quite abstractly from their significance, may be treated by the rules usually employed in those operations. . . . In this we have a most extraordinary power of signs. . . . A set of quantities, or of conditions, some of which are inconsistent with one another, are thus combined together: no idea is attached to the symbols; and from the series of operations, that may be said to be mechanical, and performed merely by the hand, a truth, applicable to quantities that really exist, emerges at last. . . . We come at truth by the help of the symbols alone; by the operations that are applicable to them alone, and that have no reference to anything really existing. Nothing, certainly, can show so clearly the power of conventional signs in matters of reasoning; and the importance, on many occasions, of neglecting the object, and attending to the sign alone.

But the line of inquiry opened by Wallis had been brought to a successful conclusion before Playfair penned these remarks. A consistent interpretation of "imaginary quantities" and the familiar operations upon them was discovered, when the former were used to represent directed lengths in a plane, "addition of imaginaries" to represent the geometric addition of such lengths, and "multiplication" to represent their rotation through designated angles.

The first to succeed in giving a complete geometric meaning to the arithmetic of imaginaries was Caspar Wessel, a Danish surveyor, whose work, however, remained unknown until after his results were rediscovered.[19] Nine years later, in 1806, two similar geometrical interpretations were published, both by Frenchmen working independently of each other. Buée, whose paper has already been mentioned, regarded the square root of negative unity not as determinate quantity, but as an addition "sign of affection" or quality which was capable of characterizing the unit of length.[20] He interpreted it as the sign of perpendicularity, and stated, in an obscure manner to be sure, definitions for adding and multiplying the directed lengths thus symbolized by the imaginaries. The other essay was by Argand, whose name is now universally associated with the representation of directed coplanar lines by complex numbers. Although his exposition of the interpretation was a model of lucidity, it was neglected for several years until Français and Gergonne recognized its importance.[21] Finally, some years later Gauss and Mourey developed such an interpretation for the ordinary complex numbers, the former in what is now its canonical form. The latter tried to introduce a new type of complex number in order to develop a calculus for the geometry of three dimensions, but he did no more than formulate the program for such an enterprise.[22]

The whole matter of the geometrical interpretation of imaginaries was discussed in an authoritative manner by John Warren. Declaring it to be a scandal that the "real nature" of imaginaries was still unknown and their "real existence" even denied, he proposed to examine and finally dispose of all the known objections against them. He pointed out that, in any case, imaginary "quantities" were capable of undergoing operations analogous to those upon ordinary quantities. He therefore concluded that "the operations of algebra were more comprehensive than the definitions and fundamental principles" of the ordinary quantities. He conceived his problem to be the finding of such "general definitions and fundamental principles" as were applicable to both "real" and "imaginary" quantities alike. And he thought he found its solution by regarding algebra as representing, in the well-known manner, directed segments in the plane.

It was the objections to this conclusion that he carefully examined. To the objection that imaginaries were merely signs of impossibility and so without "real existence," he replied that though they were signs of impossibility in some cases, they were not so in all, and that in this respect they resembled *fractions*. "If we have a question, which from its nature does not admit of a fractional answer, and in resolving this question we arrive at an equation, which only admits of fractional roots, these fractional roots are in this case a proof, that the question involves an impossibility."[23] This answer represented a great advance in thought, for it showed that "impossible" was now clearly recognized as a relative term. Whether a combination of symbols is to be declared impossible or not depends altogether on the interpretation which may be given them and upon the conditions of the problem in which they function. The question is not to be settled *antecedently* to such an interpretation. Mathematics was now well on its way to freeing itself from the shackles of a too narrowly conceived theory as to its nature.

A second objection which Warren considered was that there is no necessary connection between algebra and geometry, so that it is improper to bolster up the former by appealing to the latter. His reply was, unfortunately, much less satisfactory than to the preceding objection. Briefly, he declared that since negative as well as imaginary quantities are "mere absurdities" arithmetically considered, a convention must of necessity be made whereby such expressions represent something or other, and thereby acquire "real existence."[24] In other words, Warren regarded the algebraic operation upon imaginaries to be valid only in so far as the latter had some *specific* extrasystemic reference. This was a definite recognition that algebra *need* not have arithmetical numbers as its subject matter, and so indicated a wide departure from traditional ideas. But what if more than one distinct interpretation can be given to the symbols of algebra? What is the defense to be offered for an algebra cultivated with no reference whatever to any specific interpretation? To such queries Warren had no reply, and we must turn for an answer to a group of men who laid the basis for the modern theory of algebra and logic.

3

The view that imaginary numbers were "intelligible" when given a geometrical interpretation dispelled ancient prejudices and provided a useful calculus for the geometry of the plane. It was, however, still far from being an adequate philosophy of mathematics. It had nothing to say concerning what imaginaries were when not so interpreted, or why the mathematician could study them even apart from any of their specific interpretations. It still supported the notion that the *validity* of mathematical *demonstration* depended upon the *specific* nature of the subject matter studied, and so did not distinguish between mathematics as "pure" and mathematics as a natural science. Hence it overlooked the fact that a generalized algebra or "pure" mathematics was concerned with the *formal properties* which different subject matters shared—properties which are exhibited in the structure of the symbolism which the algebra employs. And so it failed to note that both algebra and geometry, when not cultivated as natural sciences but as "pure" mathematics, are both primarily *languages* or complex *structures of symbols;* that both can be interpreted in more than *one* way as referring to extrasystemic entities; and that both are hypothetico-deductive sciences, their task being to unravel the logical consequences of their premises, *whatever* be their extrasystemic interpretation. Such a linguistic or semantic version of mathematics was gradually developed by a group of thinkers, associated in one way or another with Cambridge.[25]

(1) *Woodhouse.* It was Robert Woodhouse who, by his penetrating remarks on the nature of symbolic reasoning, laid the basis for the subsequent development of algebra.

> Operations with impossible quantities are really regulated by the rules of a logic equally just with the logic of possible quantities. . . . If operations with any characters or signs lead to just conclusions, such operations must be true by virtue of some principle or other; and the objections against imaginary quantities ought to be directed upon the unsatisfactory explanation given of their nature and uses.[26]

Hence it was impossible to accept Playfair's weak defense of the use of imaginaries, namely, that they merely *suggested* conclusions which required to be *demonstrated* by more legitimate methods. For in what does demonstration consist anyway, he asked himself. His answer anticipates the logical developments of a century later:

> Demonstration is a method of shewing the agreement of remote ideas by a train of intermediate ideas, each agreeing with that next it; or, in other words, a method of tracing the connection between certain principles and a conclusion, by a series of intermediate and identical propositions, each proposition being connected into its next, by changing the combination of signs that present it, into another shewn to be equivalent to it.[27]

Now, reasoning with "imaginary numbers" follows exactly this procedure. For the characters denoted as "imaginary quantities" may be combined, by definition, in the same manner as the signs for "real quantities" can. Hence while the symbol $\sqrt{-1}$ does not represent a quantity, and while the "operations" of "addition" and "multiplication" upon it are not the *arithmetic* operations designated by the same names (since these latter are applicable to "real quantities" only), none the less it can be manipulated by operations as thoroughly intelligible, as much deserving the name of reasoning, as any operations whatsoever.[28]

Woodhouse was thus committed to take the inevitable step. Not only is calculation with "imaginaries" as legitimate as with the ordinary signs for quantities, but such calculation depends *in no way* upon assigning to them a geometrical interpretation. They may represent, he tells us in effect, *anything whatsoever* which conforms to the formal conditions stated in the defining properties of the algebraical operations upon them. "That, in algebraical calculation, geometrical expressions and formulas are not essentially necessary, perhaps this short and easy consideration may convince us; since algebra is an universal language, it ought merely to be competent to express the conditions belonging to any subject of inquiry; and, if adequate expressions be obtained, then there is no doubt that with such, reasoning or deduction may be carried on."[29] While therefore expressions like *sin x,* which are borrowed from geometry, occur in algebra or analysis, they denote *within* the system of algebra nothing else than certain algebraic functions of x. It is only because such geometrical *expressions* are more commodious, typographically considered, than the corresponding analytical functions, that they are employed.

(2) *Peacock.* However, it was in the writings[30] and under the influence of George Peacock that these ideas were finally crystallized into explicitly formulated principles. The distinction between mathematics as a natural science, and mathematics as the science which investigates the necessary connections between premises and conclusions, is fundamental to all his thinking. He began his famous *Report* by commenting on the difference between sciences like algebra and geometry, which are founded on assumed principles and definitions alone, and the physical sciences which do not appeal to any such *ultimate* facts as foundations. In *applying* the mathematical sciences to these latter, he held we must *interpret* the principles and definitions as referring to physical things, so that the conclusions we obtain are true only within the limits of the truth and universality of our fundamental assumptions. For the conclusions of applied mathematics are not merely asserted to be implied by the premises, they are also asserted as being *true;* they can therefore be regarded as representing only approximately the actual state of affairs.[31]

If, then, mathematics as applied is *physics,* what is mathematics not applied? What is it that requires "interpretation" before questions of material truth may be significant, but which without "interpretation" may none the less be studied with regard to the connection between premises and conclusions? Peacock has supplied important answers to these questions, without raising them explicitly.

Peacock develops his views in several stages. Arithmetic and algebra, he points out, are two *distinct* sciences, although the latter is commonly, though mistakenly, believed to be based on the former. According to the usual view, the only difference between them is that algebra employs symbols in such a way that the properties of numbers (the subject matter of arithmetic) receive a perfectly general formulation. But this notion is soon found to be inadequate, Peacock suggests, when it is seen that the fundamental operations of arithmetic must be transformed and "enlarged" radically, if they are to be employed in algebra. Because of this, a clear distinction must be made between *arithmetical* and *symbolical* algebra.

> The first of these sciences would be, properly speaking, *universal arithmetic:* its general symbols would represent numbers only; its fundamental operations, and the signs used to denote them, would have the same meaning as in common arithmetic; it would reject the *independent* use of the signs + and −, though it would recognize the common rules of their incorporation, when they were preceded by other quantities or symbols: the operation of subtraction would be *impossible* when the subtrahend was greater than the quantity from which it was required to be taken, and therefore the proper *impossible* quantities of such a science would be the *negative* quantities of *symbolical* algebra; it would reject also the consideration of the multiple values of simple roots, as well as of the negative and impossible roots of the second and higher degree: it is this species of algebra which alone can be legitimately founded upon arithmetic as its basis.[32]

Peacock thus laid the basis for distinguishing between a system of marks having an explicit *extrasystemic* reference, concerning which questions of truth and falsity are significant; and a system of marks with no such *explicit* extrasystemic interpretation, concerning which questions of truth are meaningless. And he offered an explanation of how an uninterpreted but general system of marks arises from the study of an interpreted system.

Arithmetical algebra must be distinguished from symbolical algebra, because certain operations, *always* permissible in the latter, are significant in the former only *under limiting conditions*. Thus, the expression $a-b$ is significant in arithmetical algebra only when $a > b$, and is meaningless when $a < b$, because it is not, and cannot be, defined in arithmetic. It cannot even be regarded as a "generalization" of the arithmetical relation of subtraction, since nothing within the subject matter of arithmetic corresponds to it. It is clear therefore that any theorem in arithmetical algebra involving the expression a − b cannot be asserted universally for all values of *a* and *b;* and mathematicians must put up with the great inconvenience of having to recognize exceptions to what *would* be perfectly general theorems *if* subtraction were an unrestricted operation. It seems but an easy step to allow the operation $a - b$ for all values of a and b, and mathematicians take it. But, and this is the heart of the matter, any mathematician taking it has left the *domain of arithmetic;* in strict logic, if not in imagination, he has assumed a rule of manipulating *marks or signs* whose extrasystemic interpretation cannot be integers or ratios of integers. Hence $a - b$, when a is less than b and "−" denotes arithmetical subtraction, is nonsense; it is a possible combination of signs when their extrasystemic meaning is not so restricted. *What* interpretation is to be assigned to them need not be specified, and in any case may be postponed.[33]

Nevertheless, while arithmetic does not provide the logical foundation for symbolic algebra, it has a heuristic value in *suggesting* the principles (or laws of operation) of the latter. While in one sense, the operations of symbolic algebra are arbitrarily assumed, in another sense they are not; except for the restrictions upon the range of their applicability, the operations of arithmetic obey the same *formal rules* of combination as do the algebraic ones. In arithmetic, however, Peacock holds, these rules are the necessary consequences from the definitions of integers and ratios. But when the symbols +, ×, −, ÷, no longer represent operations upon integers, these rules "become *assumptions,* which are independent of each other, and which serve to define, or rather to *interpret* the operations, when the specific nature of the symbols is known; and *which also identify the results of those operations with the corresponding results in arithmetical algebra, when the symbols are numbers and when the operations are arithmetical operations.*"[34]

The important consequences of this approach to mathematics are obvious: (1) While in arithmetical algebra (the suggesting discipline), *a, b,* etc., represent integers or ratios of integers, in the new science of symbolical algebra these signs *need not* represent integers or fractions, and in general will not be capable of being so interpreted. (2)

But it is equally true that the signs of operation, +, −, etc., which in arithmetic denote the familiar addition, subtraction, etc., of integers, in the new science are signs which cannot in general be so interpreted. Only the formal properties of the possible operations which they represent, as stated by the rules of combination of these signs, are relevant to symbolic algebra. It is also strictly immaterial whether we can in fact find any suitable extrasystemic interpretation for them.[35]

What is the outcome concerning the nature of "pure" mathematics? Peacock indicates it clearly. When formally defined operations are symbolically denoted, certain "expressions" or "forms" are obtained, which are equivalent to other forms in virtue of the rules of the system of symbols. The discovery of these equivalent forms constitutes the principal business of algebra. This is the coping stone of the view that mathematics is primarily a language, concerned with certain formal relations, which are represented in the permissible combinations of a system of signs. It defines the function of the mathematician to be the application of formulas of translation, in virtue of which expressions symbolically different are found to be equivalent.[36]

But could not a symbolic algebra be constructed independently of any suggesting science, it may be asked. Is not the function of the suggesting science merely psychological, and does not the equivalence of forms in the algebra depend upon its *own* assumed general rules of operation? Peacock considered the idea, only to reject it, because in that case "we should be altogether without any means of interpreting either our operations or their results, and the science thus formed would be one of symbols only, admitting of no application whatever."[37] This answer is far from clear; for the fact that a symbolic algebra has been constructed according to a certain method, such as deriving it from a suggesting science, does not guarantee that applications for the algebra can be found. At this point Peacock revealed a definite limitation in his outlook, a limitation, however, which was removed by other members of his school of thought. On one fundamental point, however, he saw clearly and truly: the possibility of interpreting extrasystemically the symbols of an algebra in no way affects the validity of the equivalences established by means of the formal rules of the combination of symbols which characterize that algebra.[38] It is this basic doctrine which is at the heart of modern logical theories, and which is the despair of their hostile critics.

(3) *Gregory.* The distinction between "signs of operation" and "signs of quality," already prominent in the work of Peacock, continued to characterize the Cambridge group of mathematical thinkers.[39] This point of view dominated Boole completely, a consequence of his contacts with Duncan F. Gregory and DeMorgan. While both of these men were in substantial agreement with Peacock, their writings on this subject are of extraordinary interest, since they were more critically mature in the statement of their position than he was. It was Gregory's merit to have exhibited clearly the roles of the commutative and the distributive laws. He died a young man, but he had succeeded in stating in a particularly fruitful way the principle of separating the signs of operation from other signs.

> The light in which I would consider symbolical algebra is that it is the science which treats of the combination of operations defined not by their nature, that is, by what they are or what they do but by the laws of combination to which they are subject. And as many different kinds of operations may be included in a class defined in the manner I have mentioned, whatever can be proved of the class generally, is necessarily true of all the operations included under it. This does not arise from any analogy existing in the nature of the operations, which may be totally dissimilar, but merely from the fact that they are all subject to the same laws of combination. . . . We are thus able to prove certain relations between the different classes of operations, which, when expressed between the symbols, are called algebraic theorems. And if we can show that any operations in any science are subject to the same laws of combinations, the theorems are true of these as included in the general case: Provided always that the resulting combinations are all possible in the particular operation under consideration. For it may happen, that though each of two operations in a certain science may be possible, the complex operation resulting from their combination is not equally possible. In such a case, the result is inapplicable in that science.[40]

As an illustration, suppose, for example, F and f be any two operations, whose *formal* nature is defined as follows, but about which nothing further is stated:

$$\text{(I)}\ F\,F\,(a) = F\,(a) \qquad \text{(III)}\ F\,f\,(a) = f\,(a)$$
$$\text{(II)}\ f\,f\,(a) = F\,(a) \qquad \text{(IV)}\ f\,F\,(a) = f\,(a)$$

These signs of operation may now be interpreted. If F and f denote the usual arithmetical operations of addition and subtraction respectively, then the above four relations state a part of their essential formal properties. But F and f may be interpreted differently: F may represent the translation of a point through the circumference of a circle, and f the translation through a semicircumference; this interpretation also "holds." Now while the arithmetical operations are clearly not identical with the geometrical ones, they do share certain formal properties of the latter. Hence, whatever can be proved of the symbolical operations F and f defined as above, is necessarily true of the arithmetic, geometric, or any other permissible interpretations which can be given them. It is not difficult to appreciate the fertility of the idea here sketched; and in fact the modern studies of structural invariance or isomorphism are illustrations of it.

(4) *DeMorgan*. DeMorgan, like Boole, not only made fundamental contributions to theoretical logic, but was a pioneer in establishing the nature of pure mathematics and symbolic algebra. Like others of his time, he found the usual discussions of algebra "absurdly incorrect," and devoted much effort to establishing that science on a sound basis. It soon impressed itself upon him that the unrestricted use of "operations" represented by words or other signs was an essential feature of mathematics, and that many of the difficulties in algebra were simply the consequences of a careless use of the *language* of mathematics. For words as well as other symbols, though of our own making, are treacherous things, and we often draw conclusions from one meaning of a symbol only to apply them to other, often incompatible, meanings.[41] It was particularly important in algebra, he found, to keep distinct the intrasystemic meanings of signs (their syntax, or modes of combination) from the extrasystemic interpretation which may be given them. Accordingly, he recognized two distinct parts of algebra, the technical, and the logical.

> Technical algebra is the art of using symbols under regulations which, when this part of the subject is considered independently of the other, are prescribed as the definitions of the symbols. Logical algebra is the science which investigates the method of giving meaning to the primary symbols, and of intepreting symbolical results.[42] . . . The first step to logical algebra is the separation of the laws of operation from the explanation of the symbols operated on or with.[43]

If "pure" mathematics is to be identified as a science which is free from the necessity of making material or empirical assumptions, can "pure" mathematics be recognized in either one of these distinctions? It surely cannot be the process of interpreting extrasystemically the symbols of algebra; for such a process must certainly make empirical assumptions when it interprets the symbols as representing properties found in spatial or temporal contexts. Even when the symbols are taken as standing for relational properties or "essences," like cardinal and ordinal numbers, the structure of classes, etc., material assumptions are needed at *some* point *if* these relational properties are equated with the highly abstract characters of spatio-temporally situated things. If, however, we understand by "technical" algebra not simply the actual manipulation of symbols, but also the analysis of the structure of symbolic systems, it is possible to recognize it as coinciding, in good part at least, with what is usually designated as pure mathematics. Since the difference between technical algebra (or pure mathematics) and logical algebra (or applied mathematics) is fundamental to the discussion of this paper, I shall let DeMorgan state concretely the distinction, without apologizing for the length of the quotation.

> Nothing can be clearer than the possibility of dictating the symbols with which to proceed, and the mode of using them, without any information whatever on the meaning of the former, or the purpose of the latter. A person who should learn how to put together a map of Europe dissected before the paper is pasted on, would have symbols, various shaped pieces of wood, and rules of operation, directions to put them together so as to make the edges fit, and the whole form an oblong figure. Let him go on until he can do this with any degree of expertness, and he has no consciousness of having learnt anything: but paste on the engraved paper, and he is soon made sensible that he has become master of the forms and relative situations of the European countries and seas.
>
> As soon as the idea of acquiring symbols and laws of combinations, without given meaning, has become familiar, the student has the notion of what I will call a *symbolic calculus;* which, with certain symbols and certain laws of combination, is *symbolic algebra:* an art, not a science; and an apparently useless art, except as it may afterwards furnish the grammar of a science. The proficient in a symbolic calculus would naturally demand a supply of meaning. Suppose him left without the power of obtaining it from without. His teacher is dead, and he must *invent meanings* for himself. His problem is, Given symbols and laws of combination, required meanings for the symbols of which the right to make those combinations shall be a logical consequence. He tries and succeeds; he invents a set of meanings which satisfy the conditions. Has he then supplied what his teacher would have given, if he had lived? in one particular, certainly: he has turned his *symbolic* calculus into a *significant* one. But it does not follow he has done it in the way his teacher would have taught him, had he lived. It is possible that many different meanings may, when attached to the symbols, make the rules necessary consequences. We may try this in a small way with three symbols, and one rule of connexion. Given sym-

bols M, N, +, and one sole relation of combination, namely that M + N is the same as N + M. Here is a symbolic calculus: how can it be made a significant one? In the following ways, among others. 1. M and N may be magnitudes, + the sign of addition of the second to the first. 2. M and N may be *numbers,* and + the sign of multiplying the first by the second. 3. M and N may be lines, and + the direction to make a rectangle with the antecedent for a base, and the consequent for an altitude. 4. M and N may be *men,* and + the assertion that the antecedent is the brother of the consequent. 5. M and N may be *nations,* and + the sign of the consequent having fought a battle with the antecedent.[44]

Algebra, conceived as pure mathematics, is therefore a science which studies equivalences between symbolic expressions, by virtue of their structure specified as relevant by the rules of the system. By reducing the extrasystemic references of its symbols and isolating their syntacitcal meanings, mathematics has grown from having been the science of quantity to becoming the science which explores the most abstract properties of any subject matter whatsoever.

4

No account of the growth of modern mathematics and logic can be given without including the contributions of Sir William R. Hamilton.[45] It is in large measure his work that is responsible for the freeing of symbolical algebra from the too great reliance upon arithmetic which characterized Peacock's views. Although Hamilton belongs, in a sense, to the group of thinkers discussed above, who insisted upon the indispensability of an extrasystemic interpretation of algebra, he had many points of contact with the symbolical approach. His opposition to regarding mathematics merely as a language helps to throw into relief the essential features of that philosophy of mathematics.

As in the case of his contemporaries, it was the paradoxical nature of "negative" and "imaginary quantities" which led Hamilton to the examination of the foundations of mathematics. He soon found himself in agreement with those who held that these did not represent quantities at all. But he was dissatisfied with the theory which regarded them as representing directed segments; and he could not accept for his own the linguistic analysis of their nature. What alternative besides the one already discussed remained for him?

Hamilton was a friend of Coleridge and Wordsworth and, inspired by the former, took to reading Kant. It was under the influence of the transcendental aesthetic that his views on algebra matured.[46] In a series of letters at about this time to his friend, Lord Adare, he stated his ideas clearly. In arithmetic, he explained, number is the answer to the question "How many"; in metrology, the science of measure, it is the question "How much"; but in algebra, number answers the question "How placed in a succession." Algebra, and in fact every subdivision of mathematics, is therefore the science or order, progression, or "pure Time."[47] It was possible to belong to three different schools of thought concerning the nature of mathematics: to the theoretical, the philological, or the practical, according as one chiefly aimed at clearness of thought, symmetry of expression, or ease of manipulation of symbols. Hamilton included himself in the first group "as seeking a clear and lively intuition, by whatever cost of meditation or mental discipline."[48]

On the one hand, therefore, Hamilton denied that algebra was a mere play with contentless signs; but on the other hand, he was equally insistent that no empirical or material knowledge was required to cultivate it.[49] If, however, we regard pure mathematics as dependent upon our a priori intuitions of time, mathematics will at once be a nonempirical science and also have a "content," namely "pure Time." But it was quite evident that what Hamilton meant by "pure Time," was nothing other than abstract order, or as he put it, "order in continuous progression." In any case, he was able to show in his famous essay on "Algebra as Pure Time" (by taking the notion of serial order or time as primary, and by representing "points of Time" by real numbers, "steps" or intervals of time by a "couple" or pair of numbers) how negative and imaginary "numbers" may be regarded as conventional representations of certain *relations* between pairs of numbers.[50] While Hamilton's discussion of the nature of these "steps" in time is not free from serious difficulties, his method (of considering certain relations between *sets* of two numbers each) has remained the most satisfactory approach to the analysis of fractional, negative, and imaginary "numbers" in terms of relations between integers.[51]

Hamilton subsequently turned to the investigation of an algebra of triplets, or sets of three numbers; and later, of a generalized algebra of sets of n numbers each. But he found that while the "subtraction" and "addition" of triplets could be subsumed under the formal rules for the "subtraction" and "addition" of couples, "multiplication" of triplets presented grave difficulties. Only after fifteen years of thought did he hit upon a successful rule for performing it; he was so elated at his discovery that he carved the result on a wooden bridge upon which he was passing at the time. He had been trying to find a spatial interpretation for the triplets and the operations upon them, but all those he could think of required abandoning the distributive law of multiplication; the inconveniences in calculation which resulted made this prohibitive. But now he hit upon a rule of multiplication for which the distributive law held, although it was necessary to give up the commutative principle.[52] This was the beginning of the theory of quaternions. For the first time an algebra was discovered with an indisputably useful geometrical interpretation, which illustrated a set of principles of operation radically different, even formally, from the operations of arithmetic and the algebras suggested by arithmetic. Mathematics had really come into its maturity as the science of the abstract structure of anything whatsoever.

How profound were Hamilton's differences with the linguistic school? He himself indicated that they were not far-reaching, and that his "pure Time" could be viewed as a *suggesting* science, in Peacock's sense, for constructing a symbolical algebra, whose extrasystemic interpretation could then be neglected. It was because the linguistic approach did not afford him an adequate intuition of *number* that he was compelled to dissent from it.

> I confess I do not find myself able to form a distinct *conception* of number, *without some* reference to the thought of *time,* although this reference may be of a somewhat abstract and transcendental kind. I cannot fancy myself as *counting* any set of things, without first *ordering* them, and treating them as *successive:* however arbitrary and mental (or subjective) this assumed succession may be.[53]

On this question, present discussion is still not unanimous, and the issues cannot even be presented here. One thing seems clear, however. If *pure* mathematics logically requires the notion of number which is the structure of *counted things,* then not only must the idea of *order* be taken as one of the primitive underived ideas in all mathematics (a conclusion which follows on independent grounds as well); but also what is called pure mathematics must be logically dependent upon assumptions and notions which are empirical in nature as well as in origin.

In spite of this divergence, Hamilton saw the difference between himself and the Cambridge school progressively diminish.

> I feel an increased sympathy with, and fancy that I better understand, the Philological School [he wrote his friend Robert Graves in 1846]. It enables me to see better than before the high functions of language, to trace more distinctly and more generally the influence of signs over thoughts, and to understand an answer which I hazarded some years ago to a question of yours, What did I suppose to be the *Science of Pure Kind*? namely, that I supposed it must be the *Science of Symbols*.[54]

And to Peacock he wrote the same year,

> My views respecting the nature, extent and importance of symbolical science may have approximated gradually to yours. . . . [But] I still look more and more habitually *beyond* the symbols than they [the Cambridge school] would chose to do. To use DeMorgan's image, I turn up the fronts of the bits of the dissected map, and try to learn what countries they denote, as well as how they fit together.[55]

5

It was against the background of the ideas whose growth has been sketched that Boole developed his mathematical formulation of traditional logic. He was on terms of acquaintance with the leading figures of contemporary English mathematical research.[56] Gregory was his friend, and Boole's first mathematical paper appeared in the *Cambridge Mathematical Journal* for 1840, when Gregory was its editor. Boole was in correspondence with Gregory until the latter's death, and Gregory acted as mentor to the younger man. It is not surprising, therefore, that Boole used the ideas of the Cambridge group as leading principles in his own researches.

The notion of one branch of science acting as a "suggesting science" for developing another, as well as the notion of separating signs of operation from other signs, dominated Boole's mathematical writings, as readers of his *Differential Equations* know.[57] They were also the leading principles, in slightly altered form, in the development of his logical system, as he himself acknowledged.[58] Indeed, he found that the rules of combination for signs of logical operation were the same as those of an algebra of the numbers 1 and 0; and one of the blemishes of his procedure is that he did not consistently distinguish between the development of such an abstract algebra and the interpretations which could be given to its terms. The details of Boole's algebra, as well as of his difficulties in interpreting each of its processes, are too well known to need a statement here.

It was very fortunate that the theory of symbolic algebra had been worked out far enough, so that the validity of any deduction in Boole's system was generally acknowledged *not* to depend upon the possibility of giving any specific extrasystemic interpretation to it. For, strange irony of history, Boole believed himself in his logical writings to be symbolizing the laws of mental processes.

> The design of the following treatise [he said on the opening page of *The Laws of Thought*[, is to investigate the fundamental laws of those operations of the mind by which reasoning is performed; . . . and to collect from the various elements of truth brought in view in the course of these inquiries some probable intimations concerning the nature and constitution of the human mind.

He spent much time in reading and reflecting on psychology, so that "mental science became his study and mathematics were his recreations."[59] There is at least a grain of truth in his wife's contention that Boole believed *The Laws of Thought* to be "really about brain-action"—a work on psychology, not a mathematical text; and that in studying the mathematics of the action of the brain we are studying and conversing with God.[60]

By virtue of the theory of symbolical algebra, however, it is not necessary to determine what Boole thought was *one* possible interpretation of his logical calculus. It *may* be that it expresses certain abstract aspects of the Infinite, the Absolute, and the Eternal, as his wife said; but its merits as a calculus of logic do not depend on this being or not being the case. Boole himself was clearly aware that no guarantee can be given that a symbolic calculus has a *unique* extrasystemic interpretation; there cannot be a subject matter of which it may be said that the calculus is "really about *it*" and about nothing else.

It is for this reason that Boole's work in logic is not simply a mystical excursion into transcendental metaphysics, or simply an interesting variety of an abstract algebra. For when the symbols of *The Laws of Thought* are interpreted as classes, propositions, and certain relations between them, this work becomes *their* theory, it formulates the structure of *their* relations to one another. As an uninterpreted, abstract algebra it exhibits certain analogies, in respect to the abstract structure of the operations employed, with ordinary algebra and with many other disciplines. As a calculus of classes or of propositions, although the rules of its operations may have been *suggested* by those of ordinary "quantitative" algebra or by the study of the Absolute and the Eternal, the formal similarities that it exhibits with other subject matters must be regarded as at least a contingent matter of fact.[61]

6

How contemporary discussions reflect and extend the ideas whose growth has been sketched, is too large a theme for the tail end of an historical essay. A bare reference to the writings of Peirce and Lewis in America, of the Wittgensteinized Russell in England, and of Hilbert and the logical positivists on the Continent, will perhaps suffice to indicate the fact that in current logical researches the study of the syntactical structure of signs has not diminished in importance.[62] Indeed, preoccupation of philosophers with the nature of symbolism is neither an isolated contemporary phenomenon, nor confined to the English logicians of the nineteenth century. The insights that discursive knowledge is necessarily symbolical and that language is its instrument, that the structure and rationality of the world become expressed and communicated in language, and that without systems of symbols logic can neither be nor be conceived, are hardly of so recent a date.

The theory that pure mathematics and logic are abstract calculi, may not be the last word of wisdom on the subject, but it surely should be among the first. A careful examination of its consequences shows that it makes possible a philosophy of logic able to avoid well charted reefs and whirlpools of general philosophy: a crude psychological empiricism, a romantic mental creativism, and an extreme form of Platonic realism.

1. It must by now be too evident to require many more words, that the validity of any deduction in a symbolic algebra cannot depend upon the material interpretation which may be given to the symbols of the calculus. Expressions are equivalent in a system of pure algebra, by virtue of the rules for combining the elements of the system *whatever* be the interpretation which may satisfy them. Hence the

validity of any formal deduction can never be a matter for any experimental test which appeals to some extrasystemic procedure. It is true, of course, that the operations of an abstract algebra are often so devised as to symbolize certain pervasive invariants in different subject matters. But this point raises different issues, to which reference will be made below. Meanwhile, however, the theory of pure mathematics here stated clearly must reject the naive view that the theorems of pure mathematics are simple inductions from experience.

2. It is sometimes asserted that mathematics and logic are free creations of the human mind, so that it is we who invent and build up towering intellectual systems, unchecked by anything save our fancy. Now indeed if by mathematics we understand *pure* mathematics, the construction and development of abstract symbolical structures, this is true in a large measure. We are free to select one aspect rather than another of the symbols or marks we employ as being relevant to the game of algebra we employ, just as we are free to attend to the *shapes* of chessmen and not to the material out of which they are made. We are in a measure free to decide upon how many distinct elements shall be incorporated into a symbolic calculus. We are also free in a measure to agree upon one convention rather than another for manipulating the symbols, just as we are free to decide upon what shall be the permissible moves for the knight in chess.

But this freedom is restricted in various ways. In the first place, this freedom, such as it is, is confined to building up an abstract, uninterpreted calculus. It well-nigh disappears, in any significant sense, when an algebra is "invented" which aims to represent invariant features in some subject matter. In the second place, the *material* quality of the symbols we employ itself imposes restrictions: thus, symbols confined to one plane cannot have rules of combination imposed upon them which require a three-dimensional space as a field of operation. And finally, the rules for manipulating the symbols must be consistent with one another, and with the number of distinct symbols arbitrarily assumed for the algebra. For when certain conventions are arbitrarily made about the manipulation of symbols, the consequences of these conventions are not themselves arbitrary or conventional.

3. According to a well-known view, pure mathematics and logic aim at being true in all "possible" worlds, not only in the "chance" world in which we find ourselves; and further, it is urged that so far

as pure logic is concerned, the existence of the world is an "accident." Unfortunately, "possibility," "chance," and "accident" are weasel words; and the important truth which is asserted in these dicta is not seldom translated to mean that mathematical and logical principles are imposed upon the flux of things in some sense *ab extra*. It is easy to conclude, consequently, that certain pervasive characters in nature can be deduced or derived, without any empirical study of the flux, from the fundamental principles of "pure logic." What light is thrown upon this issue by the analysis of pure mathematics as the structure of symbolic systems?

It is essential to note that when the symbols of an abstract language are given an extrasystemic interpretation, the result is *applied* mathematics. This is the case whether the interpretation makes the symbols stand for physical things and their relations, or for certain "essences" or logical entities. For the first alternative this is obvious, and, in addition, in *this* case the applied mathematics is a *natural* science. As for the second alternative, it is often forgotten that in interpreting an abstract algebra as a calculus *of* certain logical or mathematical entities, we are none the less performing an *interpretation*, which may be a mistaken one. But there is something more important: If such "essences" or logical entities are conceived as capable of being equated with the structures of empirical situations (e.g., cardinal numbers understood as being the relational structures of the familiar empirical counting processes), then material assumptions are involved in supposing them identical with certain pervasive characters found in empirical situations; and this is so even though such essences are generic traits involving a high level of abstraction. For it is usually admitted that "logical traits" represent elements of structure possessing the utmost generality possible, which are therefore pervasive in *all* subject matter.

It follows that pervasive characters of nature cannot, without a circular argument, be derived a priori from logical abstracta, since the latter are simply what we apprehend as the pervasive character of the flux. The derivation of mathematics from logic cannot therefore be interpreted as the deduction of material principles from nonempirical ones, or as the derivation of a "content" from the structure of that content. The identification of pure mathematics with logic, if it does not simply mean that both are calculi with uninterpreted symbols, can mean only that the same types of abstract relations are discoverable in the former as in the latter.

It is no mystery, therefore, that pure mathematics can so often be applied. Many branches of pure analysis, it is evident from the preceding historical survey, are developed on the basis of "suggesting sciences," whose fundamental notions, even though highly abstract, are yet drawn from observation of the natural world. It is a reasonable hypothesis that pure mathematics in general is so often applicable, because the symbolic structures it studies are all suggested by the natural structures discovered in the flux of things. Their symbolical statement and development liberates us from the press of local circumstance, and permits us to extrapolate invariant relations found in one segment of the flux to another. Santayana writes, with characteristic penetration,

> Inapplicable mathematics, we are told, is perfectly thinkable, and systematic deductions, in themselves valid, may be made from concepts which contravene the facts of perception. We may suspect, perhaps, that even these concepts are framed by analogy out of suggestions found in sense, so that some symbolic relevance or proportion is kept even in these dislocated speculations, to the matter of experience.

One parting word of caution must be said. Preoccupation with language and other symbolic structures has often gone hand in hand with the denial of the "reality" and objectivity of kinds and structures. But such a nominalism is by no means necessary to the analysis of pure mathematics here stated. *If* a calculus *does* apply, it does apply only because the structure of the system of symbols *expresses* the structure of the subject matter it deals with. Symbols and the relations between them have no magical power of being able to *represent* what does not in fact exist to be represented. Just as "knowledge is in a way identical with the objects of knowledge," so, to paraphrase Aristotle, a language which expresses truly is in a way identical with what is expressed—it is identical with it in its structure.

Whatever be the weaknesses of this theory of mathematics, therefore, its great merit lies in its insistence that the principles of any subject matter can be discovered in no other way (however such principles may be *suggested* to us) than by examining *that* subject matter. It makes the modest claim that in the procedure of pure mathematics itself may be found a hint of how to bridge the often-seeming unbridgeable chasm between the nonempirical sciences and the sciences of nature.

9 / The Formation of Modern Conceptions of Formal Logic in the Development of Geometry

Although geometry originated in the arts of land mensuration, it acquired in Greek antiquity the status of a theoretical discipline whose subject matter was not to be identified with the obvious traits of bodies and whose method was not to be confused with the empirical procedures of physical measurement. However, the substitution of a realm of unchanging forms, subsisting eternally and independently of the spatio-temporal world, for the properties of the tangible masses which geometrical studies seemed so unquestionably to explore, was not a satisfactory performance to many of those who had occasion to employ geometry in the various arts and in the pursuit of knowledge of nature. In Aristotle and Newton, to take two prominent examples, the extreme Platonic interpetation of geometry found no favor, and many lesser men shared their outlook. Nevertheless, it could not be denied that the method of geometry was not empirical, and the acceptance of the truth of geometrical theorems did not wait upon observation and experiment. That the truths of geometry are "necessary" in some sense which is not merely factual or experiential, and that geometry does formulate relations between properties of bodies, were two conclusions which seemed inescapable and at the same time difficult to reconcile. The various philosophies of classical rationalism, classical empiricism, and Kantianism, were alternative heroic attempts to establish some sort of uneasy balance between them. What is common to them all is the view that the geometry systematized by Euclid is the definitive science of space or extension.

On the other hand, it is today a commonplace in the philosophy of mathematics that the discipline designated as "pure," "abstract," "postulational," or "demonstrative" geometry does not have space or extension for its subject matter, even though that discipline may be *used* for ordering and exploring properties of bodies. As a consequence of technical developments within mathematics itself, the current view is that the "axioms" of geometry are postulates which *may* be taken as "implicitly defining" the terms contained in them—i.e., that are specified in no other way—and which in any case make no assertions about any specified subject matter. It is a fair if somewhat crude summary of the history of geometry since 1800 to say that it has led from the view that geometry is the apodeictic science of space to the conception that geometry, in so far as it is part of natural science, is a system of "conventions" or "definitions" for ordering and measuring bodies.

The object of the present essay is to trace in part the development of the shift in point of view just indicated. This change owes next to nothing to the speculations of professional philosophers and logicians, and is the outcome of technical needs and advances of mathematics proper. Nevertheless, it has had a profound influence upon modern conceptions of logic and methodology. It is plausible to suppose, therefore, that some central doctrines of contemporary logic will become illumined and made more persuasive by examining the developments which they terminate. In particular, a consideration of the procedures of mathematics within the historical settings in which they operate may provide materials for a just appraisal of the limitations of traditional conceptions of mathematics and logic, as well as of the more recent views that have replaced them.

Since the classical demonstrative geometry studies relations between such magnitudes as lengths, angles, areas, and volumes, the traditional conception of the science combined a double thesis: in the first place, because geometry explored relations of magnitude, it was taken to be an inherently quantitative science; and in the second place, because it concerned itself with magnitudes of extended objects, geometry was assumed to be the science of extension or space. The dominant contemporary philosophy of geometry rejects both of these theses. Accordingly, in the first section of the present paper some of the grounds which support the rejection of the first thesis will be briefly examined. The succeeding two sections will be devoted to a more detailed survey of the materials bearing on the second thesis; and the fourth section, finally, will consider the consequent reorientations provoked by this material to the traditional claim that geometry is an *a priori* science.

1

The doctrine that demonstrative geometry is the quantitative science of extension goes hand in hand with the allied view that arithmetic and algebra are exclusively sciences of quantity.[1] Unlike algebra, however, geometry was able to free itself from such a narrow limitation of its subject matter with comparative ease, although the introduction of algebraic methods into geometry by Descartes tended to delay this emancipation. Even in antiquity many properties of figures were studied which involved only relations of mutual position and order of figures, and did not in principle depend on relations of equality and inequality of magnitudes. The beginnings of what is now known as projective geometry, which does not employ these relations of congruence, thus go back at least to the time of Euclid, some of whose lost books apparently contained projective theorems.[2] However, the systematic investigation of geometric relations which were not quantitative but which lend themselves to as precise a study as the familiar metrical ones, remained uncultivated until the needs of the arts and technology brought them to the attention of professional mathematicians of a much later era.

The representation upon a canvas of human figures and common articles of daily need requires a technique and an appropriate set of conventions for depicting three-dimensional configurations of surfaces, lines, and points upon a plane. It is natural to find, therefore, that the needs of the Renaissance painters and architects produced an array of devices for adequately "projecting" solids upon a two-dimensional surface, with the eye as the center of projection. The problems involved require no consideration of the actual sizes of objects, but only their relative positions with respect to the eye and the canvas; and the techniques appropriate to the problems bore upon the perspective and nonmetrical properties of two- and three-dimensional configurations.[3] However, no systematic theory of perspective relations was produced, even though much material emerged for systematic study. Not until the seventeenth century was there a beginning of such a study of perspective relations. Pascal and especially

Desargues established a number of fundamental theorems in this field, and so laid the groundwork for a distinctive geometric discipline. But Desargues' work fell into neglect for a hundred years until the subject was taken up independently by Monge and his school.

A characteristic feature of Desargues' work is his use of the methods of synthetic geometry, with the consequent dependence upon visual diagrams. It is plausible to suppose that the neglect of his contributions is in part due to the fact that during this period the powerful analytic method of Descartes gained currency, and threw into the shade discoveries which seemed to depend entirely upon an apparently outmoded method. Even theorems which could be established in a simple way by the classical method of synthetic geometry were proved by means of the Cartesian procedure, and Lagrange thought it worth while to boast that in his celebrated treatise on mechanics there was to be found not a single diagram. A serious consequence of this neglect of synthetic geometry was that the ingenious innovations of engineers and designers during the period of industrialization in the eighteenth century, who were faced with problems of plane representations of spatial configurations, were left unstudied by professional mathematicians; the techniques of such representation became so many different rules of thumb connected by no comprehensive principles. Fortunately, this situation was corrected by Monge, who combined a first-hand knowledge of the problems of military engineering with remarkable mathematical powers; there awaited him a rich store of material ready for systematic treatment, and he was eminently successful in supplying the beginnings of a theoretical framework and a unified method for perspective or (as he called it) "descriptive" geometry.[4]

Into the technical details of the history of "descriptive" geometry, in which the appearance of Poncelet's treatise marks a turning point, it is not possible to enter. A host of writers contributed to its development,[5] and before the end of the first quarter of the nineteenth century the possibility of a nonmetrical geometry was explicitly recognized. For example, the distinction between the classical and the "new" geometry was stated clearly by Gergonne.

> The different theories which constitute the science of extension can be placed into two distinct classes. There are, in the first place, certain theorems which depend essentially on metrical relations found to exist be-

> tween the different parts of extension which one studies, and which, consequently, can be proved only by the aid of the principles of algebra. On the other hand, there are others which are in fact independent of these relations, and follow simply from the positions which the geometric entities upon which we reason have to one another; and although one may often deduce these latter from theorems of algebra, it is always possible, by proceeding in a suitable manner, to free them from this dependence.[6]

It is true that Gergonne was not quite accurate in this description of the content of the new geometry, though an excellent prophet of things to come. For many properties of figures which involve only perspective and projective relations were classified as metrical by the geometers of the day, for the simple reason that such properties (e.g., the anharmonic ration of four points) were defined in terms of distances and magnitudes of angles, and were therefore regarded as belonging to the domain of metric relations. The discovery that relations expressible by algebraic notations were not necessarily metrical was still far off. Indeed, it was not until 1856 that Von Staudt showed how numerical coordinates could be introduced in a purely projective way without employing ideas borrowed from the theory of congruence; and it was not until this technique was established that projective geometry, as the science of those properties of figures which remain unaltered when the figures undergo projections, could be regarded as a discipline that is completely independent of relations of congruence.[7] Nevertheless, though this point is of great technical importance, it does not seriously compromise the validity and significance of Gergonne's statement. For it expresses his clear recognition, shared by his contemporaries, that there is a distinctive subject matter for demonstrative geometry which is not metrical. Consequently, whatever might be the case with other branches of mathematics such as algebra, it was evident once for all that geometry, the science of extension, could no longer be viewed as exclusively a science of quantity.

2

This recognition of a nonquantitative subject matter for geometry was, however, only the first and easiest step in breaking away from the traditional conceptions. For whether geometry was taken to study relations of magnitude of two- or three-dimensional configurations, or whether that study was understood to include also relations of position and order, in either case the identifiable subject matter of the science was the intuitable, visualizable figures in space. And as long as geometers continued to identify their subject matter in this manner, geometry as a formal discipline could not develop that independence from the limitations of visual imagination and that unhesitating use of general methods which were even then the possession of algebra. The attempts to achieve for pure geometry the generality of analytic procedures culminated in removing this dependence upon visualization and in a radical transformation of current views on the materials of geometric study.

1) Desargues and Monge. Preliminary attempts to introduce conceptions and operations into geometry which are not limited in their application to intuitable figures, can be found in the writings of Desargues and even more pronouncedly in those of Monge and his school. Desargues proposed to treat parallel lines as a case of lines intersecting in a common point, the point of intersection, however, being removed "to infinity." And he also regarded the circle, ellipse, parabola, and hyperbola as forming a single family of curves, on the ground that all could be viewed as projections of a common figure, with the center of projection at the so-called "improper point" at infinity.[8] This "improper point," obviously, was not something to be visualized.

Monge went even further in this direction. From the point of view of his descriptive geometry, it is immaterial whether one straight line cuts another *within* a given segment of the latter, or whether the former line intersects the second on *either side* of the segment: for by simply altering the position of a body, say a cube, with respect to the horizontal plane upon which it is to be projected, the point of intersection of one edge of the body with a line in the plane would also be altered. Like Desargues, he therefore regarded parallel lines as having a common point "at infinity." Furthermore, whether two curves intersect in "real" or only in "imaginary" points was also of no consequence, because the occurrence of "real" or "imaginary" intersections depends entirely on the accidental position of the bodies to be projected with respect to one another and to the plane upon which they are projected. Consequently, Monge was led to results typified by the following example. When two circles intersect in the usual way, they have a common secant which passes through the "real" points of intersection of the circles. This line, called "the radical axis," has a number of important geometrical properties, e.g., it is the locus of the centers of the circles which cut the two given circles orthogonally. If, however, the two circles do not "really" intersect, it is still possible to determine a line with those geometrical properties, even though the line no longer passes through points of "real" intersection of the circles. It is tempting to suppose that even in this case the two circles determine a radical axis passing through their common "points of intersection"—points which happen to be "imaginary." Monge did not resist the temptation, even though the imaginary points of intersection could not, literally, be imagined.

The consequences for geometrical techniques were important, startling, and to some geometers rather disquieting. Instead of demonstrating theorems by establishing the specified property for each of the various special positions of the figures involved, as had been the custom of the ancients, general proofs were supplied to cover all cases—or more exactly, the "special" cases were no longer considered as special. As Hankel remarked, "geometry won for itself a freedom of movement, which, from the point of view of the ancient geometers, appeared like a complete lack of restraint, and which seemed to endanger to the highest degree the wonted certitude and stability of the science."[9]

2) Poncelet. However, the systematic use of "improper" and "imaginary" points was left for Poncelet, who made a serious attempt to "justify" their introduction. This came about as follows. It was a common theme for dispute among mathematicians of the day, which was the more powerful instrument for solving the current problems of geometry—the method of synthetic geometry or the method of algebra and analysis. Those who professed the superiority of the latter had the advantage of being able to point to the brilliant success of analytic methods, and the majority of contemporary geometers seemed to accept this evidence as conclusive. On the other hand, defenders of the synthetic method were led to examine the grounds of the apparent superiority of analytic procedures in geometry.

Poncelet devoted considerable thought to this question.[10] Without desiring to minimize the great powers of analysis, he set himself the task of discovering the feature of the algebraic method upon which its power depended, why geometry did in fact trail behind algebra, and whether it would be possible to employ in geometry methods which are as potent as those of algebra.[11] He began his discussion of the problem by pointing out that algebra employed abstract signs, so that as a consequence algebriac reasoning could be considered as a mechanical operation with such signs, freed from any possible reference they may have.

> Algebra employs abstract signs, it represents arbitrary magnitudes by letters which have no fixed values and which permit the magnitudes to be as undetermined as possible; consequently, algebra operates and reasons equally well on signs of non-existence as well as on signs of real quantities. If, for example, a and b represent two arbitrary quantities, in the course of operating with them it is impossible to recognize what is the order of their numerical values; and one is led, in spite of one's self, to reason on expressions like $a - b$, $\sqrt{a - b}$, etc., as if they were always arbitrary real magnitudes. The conclusions drawn must, therefore, possess this same generality, and can comprehend all possible cases, for all values of the letters which are involved. We thus obtain certain remarkable expressions, creatures of the brain (êtres de raison), which seem to be the exclusive possession of algebra.[12]

The creatures of the brain to which Poncelet referred are, of course, the negative and imaginary quantities which, at the time he was writing, were still a scandal in algebra. But scandal or not, there were few mathematicians who did not recognize their value or whose logical consciences were so tender that they would not use them; and Poncelet was justified in remarking that the power of algebra rested on its employing these creatures of the brain and on the concomitant reduction of reasoning "to a purely mechanical operation" which buttressed the memory and the attention.[13]

Is algebra, however, the sole discipline which can employ "indeterminate signs?" Poncelet thought not, and tried to show that all the advantages claimed for the algebraic method can with equal right be claimed for any science which uses abstract signs, provided that the latter are employed with that freedom from extraneous demands which is characteristic of algebra. In particular, Poncelet believed that geometry also employs abstract signs, and his first task was to show this. He pointed out that even in geometry we frequenty reason on "indeterminate magnitudes," namely, when the reasoning is "implicit." "This is what occurs," he declared, "when one is required to find certain parts of a figure which are, however, unknown both with respect to size and position; and that is why the process known to the ancients as the analytic method,[14] and so highly prized by them, did not lack the generality and power which belongs to algebra."

Why, then, if geometry does employ abstract signs, is it so much less powerful than algebra? Because in synthetic geometry, so Poncelet believed, there is a fanatically faithful and myopic dependence upon explicitly drawn diagrams. "One always reasons upon the magnitude themselves which are always real and existing, and one never draws conclusions which do not hold for the objects of sense, whether conceived in imagination or presented to sight." Indeed, we stop the investigation as soon as putative objects of study cease to have a "real" existence. Moreover, demonstrative rigor is pushed to absurd lengths: conclusions which hold for a perfectly general configuration are not immeditely extended to another configuration which differs from the first only in such inessential matters as that a point lies to the left instead of to the right of a given line, and are required instead to be supplied with an independent proof for the second figure.[15] "That is just the source of weakness of synthetic geometry," Poncelet declared, "and makes it inferior to modern, especially analytic geometry. If it only were possible to employ implicit reasoning, abstracting from the actual diagram, and if it were permissible to use the consequences of such reasoning, this state of affairs would no longer continue. Ordinary geometry, without using the notation and algorithm of algebra, would become the rival of analytic geometry, which it indeed already is whenever we have to abandon the framework of explicit reasoning."[16]

Poncelet had thus located the secret of the power of algebra: in algebra signs are employed in accordance with general rules of operation, without requiring an interpretation of those signs, at every step, in terms of the magnitudes which were originally represented by the signs. Now, is it possible to formulate a principle which would give the synthetic geometer an instrument as powerful as that possessed by the algebraist, and so permit him to proceed with equal generality? Poncelet believed it was possible, and argued as follows:

> Let us consider some geometrical diagram, its actual position being arbitrary and in a way indeterminate with respect to all the possible positions it could assume without violating the conditions which are supposed to hold between its different parts. Suppose now that we discover a property of this figure, whether it be metrical or descriptive, by means of ordinary explicit reasoning—that is, by methods alone regarded as rigorous in certain cases. Is it not clear that if, observing the given conditions, we gradually alter the original diagram by imposing a continuous but arbitrary motion on some of its parts, the discovered properties of the original diagram will still hold throughout the successive stages of the system, always provided

that we note certain alterations, such as that certain quantities vanish, etc.—alterations, however, which can easily be recognized *a priori* and by reliable rules?[17]

The question in effect contained a proposal to regard every actual diagram in geometry not as the *subject matter* of geometric study, but as a complex *sign* whose components were to be operated upon in certain ways without demanding an interpretation for them in terms of anything visualizable. Poncelet was too much steeped in the traditional conception of geometry, even though he was struggling to emancipate himself from it, to see the matter in just this way. He was content to be able to answer his question in the affirmative. He argued that such an answer was consequent upon employing implict reasoning, and supported his contention by calling attention to the use of similar procedures by Carnot[18] and other mathematicians who contributed to the foundations of geometry, mechanics, and the infinitesimal calculus.[19] Poncelet was in fact simply extending the "principle of analogy" employed by Kepler in the latter's study of the interrelations of the various conic sections,[20] and his own formulation of the principle which governed the procedures he was proposing to use would no doubt have given Kepler much pleasure.

It is at bottom simply the *principle of permanence or indefinite continuity of mathematical laws with respect to quantities varying insensibly,* a continuity which for certain states of a given system often exists only in a purely abstract and ideal manner. . . . The principle of continuity, considered simply from the point of view of geometry, consists in this, that if we suppose a given figure to change its position by having its points undergo a continuous motion without violating the conditions initially assumed to hold between them, the . . . properies which hold for the first position of the figure still hold in a generalized form for all the derived figures . . .[21]

There was only one difficulty in using this principle to which Poncelet was sensitive, namely, how to differentiate a "general" or "indeterminate" position of a geometrical figure from a special or particular one. He believed, however, that the distinction involved could always be established in pactice for a given problem.[22] On the other hand, he found no difficulty in the fact that a thoroughgoing application of the principle led to apparently paradoxical results, such as that two parallel lines would have a common point "at infinity" or that two circle would always have two common points even when they do not "really" intersect. Indeed, while Poncelet believed that his principle had been used by geometers long before he formulated it, he regarded it a serious flaw in their procedure that they had stopped short at introducing various "imaginary" elements into geometry and had restricted the application of the principle so as not to leave the bounds of "real" or "actual" configurations. In short, the "paradoxes" were no obstacle, for such "paradoxes" also occur in algebra. "Is it reasonable to reject from geometry ideas which are universally admitted in algebra, whose rigor is exceeded by no science? Have we not already admitted into geometry the infinitely small and the infinitely great, though their existence is purely hypothetical? What prevents us from accepting the imaginaries as well?"[23]

It is easy to carry away the impression that Poncelet thought of his principle primarily as a heuristic device, valuable because it helped the discovery of interesting geometrical properties and because it introduced a powerful unifying method into the sciences. He was, however, keenly concerned with the status of the principle of continuity as a principle of proof. "What harm will result," he asks, "especially if we are rigorous in our proofs, if we do not content ourselves with half-proofs, and if we reject analogy and induction, which are frequently misleading and are not to be confounded with the principle of continuity?" For analogy and induction lead us to assert universal conclusions from isolated particular facts with no necessary relations between them. The principle of continuity, on the other hand, demands that we reason from certain general conditions imposed upon a figure and that we never replace these conditions by others more general; it also demands that these conditions be fulfilled throughout the series of gradual transformations to which the figure is subjected, so that while parts of it may move off to infinity and others vanish completely, the initial general conditions are never violated.[24] Poncelet did not doubt therefore that his principle was as valid a rule for drawing geometrical conclusions as is Aristotle's *dictum* for the classical syllogism.

To appraise Poncelet's own estimate of his principle it is imperative to determine what Poncelet accomplished by introducing it. The cardinal point to be stressed is that while *nominally* he simply gained a new instrument for the well-known science of extension, *in fact* he radically transformed that science. In maintaining that a geometrical diagram is simply an "abstract sign" (or variable), subject to certain general conditions of operation, exactly as algebraic diagrams (i.e., letters of the alphabet, etc.) are variables subject to certain different rules of combination but which could be left uninterpreted, he was rejecting in so many words the idea that it is the visible figures which constitute the subject matter of geometry. And in maintaining that operations upon geometric diagrams, heretofore restricted so as always to yield actual points of intersection, were to be performed without such a hampering qualification, he was in effect proposing a *new* geometrical calculus which happened to resemble the older geometry in certain ways. It is true that Poncelet was not conscious of what he had done, and he did not in fact possess an adequate philosophy of symbolism in terms of which he could have viewed his accomplishment in an appropriate light. He would hardly have assented to the view that the task of the pure geometer (or the pure mathematician in general) is the exploration of the mutual interrelation of signs governed by specified rules of operation, irrespective of the "intepretations" or "meanings" which may be assigned to them. Nonetheless, at least on one occasion he was definitely on the road which terminates in such an outlook. "I agree with you," he wrote to Terquem, in reply to the latter's comments on the principle of continuity, "that algebra is not simply the science of quantity, but the science of algorithmic operations. But I do not conclude that there are in algebra *two kinds of impossibility, one metrical and the other algorithmic.* On the contrary, I say that there is only one kind, namely metrical impossibility."

In illustration, he pointed out that if we wish to decompose $a^2 + b^2$ into two linear factors, this is not possible as long as we interpret this expression and its supposed factors as numerical magnitudes: the operation is thus numerically impossible; but the problem is not impossible *algorithmetically,* since in fact we do have

$$a^2 + b^2 = (a + bi)(a - bi).$$

He also pointed out that in the analytic equation

$$\log(-1) = (2m + 1)\pi i,$$

we must regard "log(− 1)" as a sign, as "an expression which is in no way illusory," and the entire equation as something perfectly legitimate—though difficulties naturally will arise if we insist on interpreting it in the manner in which we usually interpret the sign "log(1)."[25] Consequently, the principle of continuity, in effect if not in explicit intent, simply was a proposal to treat geometrical diagrams and geometrical terms as *variables*, tied down to no specific interpretation, so that their mutual interrelations could be legitimately studied even if no visualizable references for them could be specified.

The consistent application of these remarks to the geometry reconstituted by the principle of continuity would have led Poncelet to anticipate insights into logical techniques which were not won until relatively recently. As it is, his views remained unclear on essential points. It does not seem to have occurred to him that when, on the authority of the principle of continuity, a "straight line" *always* has two "points of intersection with a circle"—even when the *usual* sort of straight line does not intersect the *usual* sort of circle—a change had occurred in the subject matter of geometrical inquiry as well as in the chracter of that inquiry. The familiar sort of line does not in fact always have points in common with every circle of the familiar kind. Consequently, the principle of continuity, instead of enunciating a new "fact" about the familiar lines and circles, simply aids in the *construction* of a new system of geometry which happens to employ familiar terms, though in an altered way. These terms are "implicitly defined" by the new sytem; i.e., the new system specifies the possible operations into which these terms may enter within the system, but does not assign a specific reference to them outside the system. The specification of the possible operations is made in terms of relations whose formal properties alone are given though it happens that relations with those formal properties have been *suggested* by the materials of investigation of geometry considered as the science of *extension*. Thus, a line and a circle, in the ordinary sense of these words, do not always intersect, and the principle of continuity can not alter this fact by simply postulating a new kind of "point" which is to be on par with what is *usually* designated as a "point."

Hence the principle of continuity postulates a new domain of entities, implicitly defined by the new system of geometry, which neither includes nor is included in the original one: in this new domain certain things called "lines" and other things called "circles" will always have something called two "points of intersection" in common. And it so happens that in this new domain, whose character is specified in no other way than by the postulated formal properties of its elements and relations, the relations between its various elements resemble in certain obvious ways the relations between the elements in the original visual field, though without being *identical* with the latter. In brief, therefore, the principle of continuity is primarily a device for constructing a formal system of signs on the basis of suggestions received from a system already constructed and even interpreted. Just as the "imaginary numbers" in algebra are not an additional kind of numerical magnitude homogeneous with the familiar cardinals, but are simply signs "implicitly defined" by the rules of operation into which they enter, so the "imaginary points" generated by the principle of continuity are also uninterpreted signs for something or other, whose range of denotation is limited only by the rule of combination specified for them.

Poncelet may therefore be credited with introducing into geometry a method of "extending" or "generalizing" a system of formal operations which is similar in all respects to the principle formulated for algebra by Peacock and which was baptized by Hankel as the "Principle of the Permanence of Formal Laws."[26] Poncelet saw clearly that the justification for introducing ideal and imaginary elements into geometry is no different than the principle which justifies their introduction into algebra. Fractional, negative, and imaginary "numbers" each in turn permitted the extension of the field of operations employed in arithmetic, and the principle of continuity of formal operations which directed these extensions was therefore an obviously fruitful heuristic principle in algebra; why not employ, therefore, an analogous principle in geometry?[27] As he shrewdly remarked,

> It is not the explicit introduction of the creatures of the brain (*êtres de raison*) which has led to the principle of extending the formulae of analysis; on the contrary, the implicit use of this principle has led to these creatures of the brain and in general to all the metaphysical notions which necessarily follow from it: these notions were rejected or misunderstood by the ancients, but in our day they have become so familiar that we have as much confidence in them as in truths generally admitted as proved in the most rigorous manner.[28]

Just such an extension of the *calculus of geometry* was to be achieved by the principle of continuity. Different geometrical magnitudes would be represented by letters of the alphabet, though the absolute magnitudes would not, so that their mutual relations could be investigated in abstraction from the specific configurations to which they can be made to apply. Even when the "configurations" cannot be constructed "geometrically," they may nevertheless still be regarded as "existing," precisely because no interpretation in terms of actual visualizable figures is required for the "abstract signs" of the geometrical calculus whose formal interconnections are unambiguously determined.[29] The uninterpretability of the abstract geometrical system thus obtained in terms of visualizable configurations was regarded by Poncelet to be no more an objection to such a system than is the uninterpretability of all the formulae in analysis in terms of the simple quantitative relations of elementary arithmetic. Poncelet was fond of repeating D'Alembert's pragmatic maxim: "Allez en avant, et la foi vous viendra."[30]

Poncelet's defense of his principle of continuity was formulated far too unclearly and carried the seeds of a philosophy of mathematics far too revolutionary to persuade all his contemporaries. The principle became the subject of a long controversy, which began with Cauchy's review of one of Poncelet's monographs.

> M. Poncelet employs in his *Memoire* [Cauchy remarked] what he calls the principle of continuity. As far as geometry is concerned, this principle supposes that when a configuration composed of a system of lines or curves possesses certain constant properties, although the absolute or relative dimensions of its different parts may vary in some arbitrary manner between certain limits, these properties will necessarily hold when the dimensions pass outside the limits originally set for them; and that, if some parts of the figure vanish on this last assumption, the remaining parts continue to possess the properties of the original configuration. *To speak strictly, the principle is only a bold induction,* with whose aid theorems already proved may be generalized, in the first place under certain specified conditions, and secondly even when these restrictions are lifted. Applied to curves of second degree, the principle leads the author to precise conclusions. Nevertheless, we think the principle can not be accepted in general and applied without reservation to all kinds of problems in geometry or even analysis. It is well known, for example, that in evaluating definite integrals, that is to say lengths, areas and volumes, there exist a great number of formulae which no longer are true when the values of its arguments fall outside certain limits.[31]

Remarks such as these are clearly in order if the principle of continuity is taken as a principle of inference on the strength of which one may conclude that a property, found to obtain when the parts of a figure satisfy certain initial conditions, also holds when those conditions are no longer satisfied. For example, during the eighteenth century many mathematicians committed serious errors when they assumed that because an infinite series converged for *certain* of the possible values of its variables, it converged for *any* of the possible values. Cauchy was just the person to bear in mind some of the minor scandals which resulted from neglecting the conditions of convergence, and Poncelet's formulation of his principle not unnaturally gave the impression that he was indeed proposing it as an inferential principle of the sort indicated. When understood in this sense, nothing that Poncelet said in defense of his principle can be taken as establishing it on a firm foundation.

However, if the principle of continuity is introduced as a *guide* for extending the range of geometrical operations and thereby for constructing a new "geometrical" calculus, Cauchy's remarks are beside the point. Poncelet no doubt intended the principle to "justify" the introduction of certain *expressions* which cannot be interpreted on the traditional views concerning the subject matter of geometry. A "justification" it certainly was not. As has been indicated, the function of the principle was that of a heuristic rule for developing a new symbolic system on the basis of suggestions derived from a system already constructed and familiar, and whose operations were restricted in definite ways. Poncelet seemed to believe that the new system was simply a logical consequence of the older one, or that it was an "extension" and "generalization" of the latter in such a way that the new geometry included the traditional geometry as a special case. This is both a mistaken and misleading view. The new system, constructed on the basis of hints offered by the principle of continuity, was in fact a *distinct* discipline, with laws of operation and combination formally analogous in many ways to the laws of operation in classical geometry, but logically independent of the latter.

In spite of Poncelet's fundamental unclarity concerning the import of his principle, he was nevertheless frequently aware of the *postulational* role which it played in the supposed "generalization" which it effected for geometry as the science of extension.[32] For it was simply by postulation that the ideal and imaginary elements of algebra were introduced into geometry; and Hankel was just in his comment that when Poncelet proposed to "generalize" geometric proofs and theorems with the aid of his principle, geometry was receiving a gift from analysis.[33] Since the point is of some importance, it is worth while to state the issues at greater length. From the point of view of the accepted subject matter of the then contemporary geometry, "imaginary points," "imaginary lines," etc. could be viewed as nothing other than paradoxical and absurd, just as "imaginary numbers" are strictly speaking nonsense when the subject matter of algebra is taken to be the cardinals and their familiar relations.

Now imaginary numbers can be introduced into algebra in at least three ways: by postulating a new kind of entity designated as "imaginary number" which happens to satisfy most of the operations valid for the usual sort of number, i.e., by the so-called method of postulation; by introducing a new *symbol* such as "$\sqrt{-1}$" or "i", in accordance with explicitly stated rules of operation which "implicitly define" the range of possible application of the symbol, but specify the sort of things which *may be denoted* by it in no other way, i.e., by the so-called method of implicit definition; and finally, by employing the term "imaginary number" as a short-hand symbol for certain *pairs* of integers, fractions, and other real numbers, when the operations upon such pairs are governed by stated rules, i.e., by the so-called method of construction. Historically, imaginary numbers were introduced into algebra via the method of postulation, and were looked upon with suspicion because there were no good reasons advanced that there are such things; but while their admissibility and intelligibility were challenged, new symbols were being employed in accordance with the second method, although this was not recognized at the time; and the controversy came to an end only after Hamilton developed his technique for " constructing" the imaginaries out of pairs of real numbers, in accordance with the third method.

Now the situation in geometry with respect to the imaginary and ideal elements was precisely analogous. Ponceloet and his forerunners introduced them by the method of postulation. The principle of continuity was a half-obscure device for introducing them by the method of implicit definition, although Poncelet did not clearly see this, and instead of recognizing that he was in fact constructing a new calculus, imagined that his principle supported the view that such terms as "imaginary point" *denoted something* in the subject matter of *traditional* geometry, on par but not identical with what is usually denoted by "point." Such a supposition was natural though mistaken. In order to work it out consistently, Poncelet should have offered a geometrical interpretation (in terms of actual configurations and their relations) of the ideal and imaginary elements which he postulated.[34] He did not do so, but had he done so he would have had to employ a technique similar to the one employed by Hamilton. What Poncelet left undone was supplied by geometers to be considered subsequently, who "constructed" the imaginary elements out of the usual sort of elements and their relations, and so established a "generalized" geometry on a basis independent of algebra.

Nevertheless, Poncelet accomplished an important revolution: he showed that it was possible to develop a pure geometry which need not be interpreted in terms of the intuitable subject matter of the traditional science, and although he left the matter in a logically unsatisfactory state, his work supplied powerful arguments for the view that pure geometry does not have spatial relations for its subject matter.[35]

3) Gergonne. The second method mentioned above for introducing imaginary elements, which is commonly designated as the method of "implicit definition," is a procedure which plays an important role in modern logical theory. Although the conscious use of such "definitions" is relatively recent, the essential point in its technique was grasped by Gergonne, the influential editor of the *Annales de Mathématiques* and contemporary of Poncelet; before illustrating the application of such definitions within geometry it will be of some interest to consider his views.

Gergonne began his essay on definitions with a sketch of the natural history of languages. He explained that words are used for things, for collections of things, and for abstract properties of things which are constant for a large range of objects. Gergonne's opinions on many relevant topics reflect the sensationalistic theory of abstraction common to the British empiricists and Condillac:

> We are led to detach these common properties from the objects in which they inhere . . . and assign names to them. We thus get words representing colors, odors, tastes, etc. These words designate neither individuals nor collections of individuals, but only the common way they affect us. But since originally it is things which are represented by words, we are led to suppose that all words express things which exist independently and actually. So one talks of the circularity of a figure and asks what happens to the circularity of a ball of wire when the wire is used up. Such realities, mistakenly attributed to pure conceptions of the mind, have caused vain disputes. For in creating the word *white* or *whiteness,* we have done nothing else than unite into one and the same class all objects in which this color is exhibited.[36]

In addition to the names already mentioned, Gergonne points out that an adequate catalogue of terms which are essential for a language must also contain words representing relations, such as "above," "equality," and many others.

Because of new needs the vocabulary of a language will from time to time require some kind of extension, so that new words will have to be introduced. According to Gergonne, the new words simply serve as convenient abbreviations for complex functions of terms already present in the language in question. Such abbreviations are inescapable practically, though apparently dispensable theoretically, because we are limited psychologically and require a compact notation if we are to grasp the sense of complicated ideas. Thus, it is convenient to replace the expression

$$x - \frac{x^3}{3!} + - \frac{x^5}{5!} \ldots$$

by "sin x," since with the help of the latter notation we can apprehend more readily and clearly the various properties of the series.[37] Definitions of this sort Gergonne calls *"explicit"*: they are alleged to be conventions which establish "an identity between the meaning of two expressions, of which the simpler is new and arbitrary, while the other, more complex, consists of words whose sense is already fixed either by usage or prior convention."[38] Explicit definitions are always definitions of *words,* not *things,* they are arbitrary or conventional, and they are theoretically dispensable because the defined term may always be replaced in any context by the defining expression.[39]

However, explicit definitions are not the only way in which new terms can be introduced into a language; and it is not possible to define all words in this way if we wish to avoid circularity. For example, the names of individuals cannot be defined explicitly. But how then do we come to understand the meanings of such words? In the case of new words representing sensible qualities, the presentation of sense objects will adequately introduce and specify the terms. In the case of words like "desire," "believe," "relation," and "above," however, which according to Gergonne represent intellectual simples, their meaning can be obtained only by attending to the circumstances in which the words are employed. In general, this is the method which must be employed for discovering the sense of a word when a number of other words in the language are already understood. Expressions which reveal the sense of certain words they contain by means of the understood sense of *other* words they also contain. Gergonne calls the "implicit" definitions of the words in question.[40]

Gergonne throws an interesting light upon implicit definitions when he compares explicit and implicit definitions with equations that are solved and unsolved respectively. Thus a sign "x" may be defined explicitly as an abbreviation for "1 + 1"; it may also be defined implicitly as the root of the equation

$$x^2 - 4x + 4 = 0.$$

In the latter case, it is assumed that the sense of the various signs other than "x" is already understood, and that the various operations upon numbers have been formulated. This illustration is trivial. It may even be misleading in so far as it suggests that every implicit definition of a term may be "solved" as to yield an explicit definition of it. Nevertheless, it helps focus attention on the essential point involved, namely, that an expression can be assigned a "meaning," not by defining it nominally or even ostensively, but by specifying in detail the relations into which it enters with other expressions, the rules of operation with such expressions, and the conditions of application to which these are subject. The bearing of this method of definition upon the introduction of imaginary elements into geometry has already been mentioned. The subsequent history of pure geometry repeatedly illustrates how the use of this method has compelled mathematicians to abandon the traditional conception of the science, and to espouse the view that pure geometry is a symbolic system whose development does not wait upon establishing a specific reference for the terms defined only implicitly by the system containing them.

4) Grassmann and Von Staudt. Of the three methods mentioned above for introducing imaginary elements into geometry, the method of postulation is the least satisfactory and is open to the gravest logical objections. Each of the remaining methods have been employed repeatedly, alternatively, and independently by mathematicians of the nineteenth century. Poncelet's principle of continuity represented an ambiguous compromise between the methods of postulation and of implicit definition. It is worth while therefore to consider illustrations

on the clarified use of the methods of implicit definition and construction, through which the range of application of geometry became enlarged and the philosophy of mathematics became radically altered.

a) Grassmann's *Ausdehnungslehre* of 1844 is a brilliant though obscurely written illustration of the use of the method of implicit definition. The germinating idea for the work was formed in its author's mind in connection with his reflections upon the meaning of negative signs in geometry. He noticed that if A, B, C, are any three points on a line and AB, BC, and AC the distances between them, it is possible to assert that AB + BC = AC, no matter whether B lies between A and C or not. If, however, B does not lie between A and C, Grassmann pointed out that "AB," "BC," and "AC" cannot be interpreted simply as measures of distances: they must be interpreted as denoting distances associated with a *direction*.[41] It was essential, therefore, to distinguish between the sum of lengths and the "sum" of directed lengths. And it occurred to him to consider more closely the latter kind of "sum," not only when the associated directions were identical or opposed, but also when distances were associated with any direction whatsoever.[42]

These elementary reflections became the stimulus for the construction of what Grassman believed was a new science, the "general science of pure forms," considered in abstraction from every interpretation of the forms in terms of some intuitive content. In consequence, Grassmann was led to distinguish between "formal" and "real" sciences—between pure mathematics as the science of forms in general, and applied mathematics such as geometry conceived as the science of space. Why was Grassmann forced to make such a distinction? The details of the *Ausdehnungslehre* are the best answer to this question, but Grassmann's preliminary discussions are already sufficiently revealing. "Geometry is not to be regarded as a branch of mathematics," he declared, "in the sense that arithmetic or the theory of combinations are. For geometry refers to something given by nature (namely space) and accordingly there must be a branch of mathematics which develops in an autonomous and abstract way laws which geometry predicates of space." In this division of mathematics, "all axioms expressing spatial intuitions would be entirely lacking. The foundations of this science would thereby become as evident as those of arithmetic, which at the same time the restriction that it be limited to the study of a three-dimensional manifold would also be dropped."[43]

These dicta are somewhat clarified by Grassmann's further comments on the distinction between formal and real sciences. "Proofs in formal sciences do not go outside the domain of thought into some other domain, but remain completely within the field of combinations of different acts of thought. Consequently, the formal sciences must not take their point of departure from *axioms,* as do the real sciences, but will take *definitions* instead as their foundation."[44] Formal sciences are characterized by the fact that their sole principles of procedure are the rules of logic as well as by the further fact that their theorems are not "about" some phase of the existing world but are "about" whatever is *postulated* by thought.[45] "Pure mathematics is the science of *specialized being* generated by thought. Specialized being so understood will be called a form of thought or simply a form, so that pure mathematics is the doctrine of forms." In consequence, the designation "science of quantity" for mathematics is a misnomer: for that designation would exclude the general theory of combinations from mathematics, and it applies even to arithmetic only in an improper sense. Instead of being a branch of pure mathematics, geometry (i.e., the science of space), like kinematics and mechanics, is simply an *application* of pure mathematics to the real, sensible world.[46]

It is evident from these remarks of Grassmann that whatever else his general science of forms may be it cannot be identified with a natural science. One may expect, therefore, that questions concerning the appropriateness of its definitions or the truth of its propositions—in a sense other than that of "logical truth" or validity—were regarded by him as irrelevant. This expectation is confirmed by an analysis of the detailed procedure of the *Ausdehnungslehre,* or theory of extensive magnitudes.[47] Grassmann began his technical account by introducing a dyadic "operation" upon "elements," which he symbolized by "$\frown$" and called the "synthetic connective." It is noteworthy that the synthetic connective is characterized exclusively in terms of carefully stipulated formal properties: we are told nothing further than that it is associative and commutative. Next, an "inverse operation" to the synthetic connective is introduced, and symbolized by "$\smile$" and designated as the "analytic connective." This second operation is also characterized simply by the formal properties stipulated for it, and the interconnections of the analytic and synthetic connectives are discussed in detail solely on the basis of such properties imputed to them. It turns out that the synthetic connective possesses all the formal traits of algebraic addition, while the analytic connective has those of algebraic substraction. This fact, however, was regarded by Grassmann as being strictly irrelevant to his immediate aim, though it suggests the *source* of the formal traits which he assumed for his two connectives and also indicates possible ways of interpreting them. In exactly the same way, Grassmann introduced another operation, symbolized by " $\frown$ " and designated as "multiplication." The reason for this name is that $\frown$ has *some* of the formal properties of algebraic multiplication: it is distributive with respect to both the synthetic and analytic connectives. No further formal traits are assigned to $\frown$, so that it is a more "general" operation than algebraic multiplication. Subsequently Grassmann distinguished different species of his general "multiplication" operation (as many as sixteen), by introducing different conditions which each must satisfy: for example, some are commutative, others are not.[48]

The remarkable thing is that Grassmann developed the whole of the *Ausdehnungslehre* in just this way—in terms of "elements" and "operations" which are characterized simply by the stipulated formal relations into which they enter. Accordingly, an element is to be considered simply as some specialization of form, distinct from other specializations. "No meaning is assigned to an element other than that. It is completely irrelevant what sort of specialization an element really is, for it is simply a specialization with no real content; it is also irrelevant in what respect one element differs from another, for it is specified simply as being different, without assigning a real content to the difference."[49] When further elements are later introduced, the new elements are to be distinguished from the previous ones simply on the basis of the abstractly formulated operations which are to be performed upon them. Something of the content of the *Ausdehnungslehre* may be gleaned by comparing certain of its procedures with analogous "constructions" in ordinary geometry. In geometry, a point undergoing continuous motion is taken to generate a line, a moving line is considered to produce a surface, and so on. In the *Ausdehnungslehre,* what are called "alterations" of elements determine extensive structures of the first level, "alterations" of first level structures determine structures of the second level, and so on *indefinitely*.[50] But these "alterations" and "structures" of different levels are not required to have an interpretation in terms of some intuitive content, though such interpretations *may* be found subsequently.

> My *Ausdehnungslehre* [Grassman explained in 1877] is the abstract foundation for the doctrine of space (geometry), i.e., it is free from all spatial intuition, and is a purely mathematical discipline whose application to space yields the science of space. This latter science, since it refers to something given in nature (i.e., space) is no branch of mathematics, but is an application of mathematics to nature. . . . A branch of mathematics is necessary in which the idea of a continuously varying magnitude comprehends the notion of differences which correspond to the dimensions of space. My *Ausdehnungslehre* is this branch of mathematics. However, its theorems are not simply the formulation in an abstract language of the propositions of geometry. For while geometry is limited to the three dimensions of space, this abstract science knows no such limitations.[51]

Further details of Grassmann's technical achievements may be omitted. Viewed algebraically, his work consists in developing the theory of higher complex numbers, with applications to geometry and mechanics.[52] Enough has been said, however, to indicate his general outlook and the essentials of his method. Especially important for illumining the changing conceptions of mathematics are the following: In striving for generality of formulation, he so expanded the calculus of traditional geometry that his formal system could no longer be conceived as dealing specifically with "space," and he was compelled by the very nature of his results to distinguish sharply between pure and applied mathematics. Pure mathematics was to be developed entirely with the help of logic alone. Moreover, in order to proceed formally, without a reference to the contents of any science of nature, he recognized that he could not build his system upon axioms (i.e., some set of true *propositions*); and required instead what he called "definitions": these "definitions" implicitly defined the various elements and operations which he introduced, so that Grassmann was one of the first mathematicians who explicitly recognized that mathematics is concerned exclusively with formal structures.

Partly because Grassmann interlined his treatise with awkwardly expressed philosophic beliefs, and partly because the contemporary scene was dominated by Kantian views on the indispensability of intuition for mathematics, his work was ignored by both mathematicians and philosophers. For example, Apelt, who was then professor of philosophy at Jena, expressed himself as follows in a letter to the mathematician Moebius: "Have you read Grassmann's remarkable *Ausdehnungslehre*? It seems to me that a false philosophy of mathematics is at the bottom of it. The essential character of mathematical knowledge, that it is intuitive, seems to be excluded from it completely. An abstract theory of extension such as Grassmann wishes, can be developed only from concepts; but the source of mathematical knowledge is found not in concepts, but in intuition." Moebius had similar convictions, and in his reply to Apelt declared he had not been able to read much in Grassmann's book because "it keeps itself too much aloof from all intuition, which is the essential trait of mathematical knowledge."[53] But the novelty and the consequent difficulty of Grassmann's work lay only in this, that he did not refer to any specific subject matter in his investigation: what the subject matter of the *Ausdehnungslehre* is can be gathered only from the implicit definitions of the terms which occur in it. While the "suggestive science" which stimulated the construction of the *Ausdehnungslehre* was undoubtedly projective geometry, it did not have the familiar configurations of geometry for its subject matter: the new abstract science was "about" *anything* which could be a valid interpretation for its implicitly defined terms.

b) Two techniques for enlarging the classical system of geometry have now been illustrated: the logically unsatisfactory method of postulating imaginary elements, in the manner of Monge and Poncelet; and the method of implicitly defining terms by specifying the formal relations into which they enter, in the manner of Grassmann. It remains to say something about the method of construction. This third method was apprehended, if only dimly, by Chasles, but was brought to perfection within geometry by Von Staudt. However, the details of Von Staudt's interpretation of "imaginary elements" can be fully grasped only if familiarity with the technicalities of projective geometry is presupposed; a bare sketch of his procedure will therefore have to suffice.

Von Staudt saw clearly that the so-called "generalization" of familiar concepts in mathematics as well as the introduction of new ones are usually undertaken for the sake of eliminating inconvenient exceptions to what would otherwise be perfectly general theorems in some branch of the subject. He pointed out that imaginary elements had been introduced into analysis and geometry just for this reason. They had simply been postulated, so that certain "operations" which heretofore could be performed only under certain restrictions were capable of being performed without such qualifications. But the method of postulation was, for Von Staudt, a very unsatisfactory device, since its use raised logical issues which it would not settle. As he put it,

> A point is said to be imaginary in analytic geometry when its coordinates are not all real numbers. But in this way it is only the *language* of algebra which is applied to geometry and one has not established the fact that an imaginary point, like a real point, is something independent of the coordinate system. Everyone asks quite justly, where is an imaginary point if we abstract it from the coordinate system? [54]

The researches contained in the second of Von Staudt's important essays on projective geometry are undertaken with the intent of answering such a question.

Stated more concretely, the object of Von Staudt's study was the specification of certain geometrical relations between "real" or "actual" elements in the ostensible subject matter of geometry, so that these relations, whose "existence" was in no way in doubt, could be taken as the matters designated by the expression "imaginary elements." Now such a specification could not be completely arbitrary, although at first sight it appears as if it could. For when imaginary elements were introduced into geometry by simply postulating their "existence," certain very definite relations between imaginary and real elements were postulated at the same time—indeed, the preservation of these relations under all possible arrangements of the "real" configurations of geometry was the *raison d'être* for introducing the imaginaries. Consequently, the relations between "real" elements which Von Staudt sought as the referents for "imaginary elements," were required to stand in the *same* formal relations to one another and to the "real" elements as did the postulated imaginary entities. For example, if three planes are supposed always to determine one point, irrespective of whether the planes and points are real or imaginary, then the only "real" referents for the "imaginary plane" and "imaginary point" which may be specified are those for which it would be correct to say that three "real" relations of the first kind always "determine" just one "real" relation of the second kind. In brief, the structure of the system of "real" referents for "imaginary elements" must be formally identical with the structure of the system containing real and *postulated* imaginary elements.

Von Staudt was able to specify relations between "real" configurations which satisfied this important condition. Omitting all technical details, his proposal was this: Projective geometry studies certain relations between "real" points situated on "real" lines; and when the points on a line are correlated with one another in a specified manner, they are said to stand to one another in a relation called "involution." And Von Staudt showed that if we interpret the phrase "imaginary point" to refer to involutions of real points on real lines, then all the theorems which are nominally about imaginary points become true statements about involutions. Consequently, it becomes unnecessary to postulate the existence of something called "imaginary points" which would be distinct from the elements and relations in the initial domain of real points; for the expression "imaginary point" can be taken as a manner of speech, a shorthand device, for making statements about sets of real points related in well-established ways. Von Staudt offered analogous interpretations for the expressions "imaginary line" and "imaginary plane." He also showed how statements of the form "an imaginary point lies on a real line" may be construed as elliptical statements about real points and real lines. The net outcome of his researches was that every theorem in projective geometry which was ostensibly about imaginary elements or about imaginary and real elements, could be translated into demonstrable theorems about involutions of real elements.[55]

Von Staudt's studies are therefore not only an important technical contribution to projective geometry, but also illustrate a significant logical technique. On the usual meaning attached to the expression "point" (the meaning which assigns to points some sort of sensory status), there are clearly no "imaginary" points in the subject-matter of geometry, viewed as the science of extension. Von Staudt showed, however, that certain relations between or operations upon "genuine" points may be selected in such a way that *complexes* of these relations or operations are related with one another in a manner essentially similar to the way the "genuine" points are related to one another. In other words, he showed how well-defined configurations with well-defined relations between them may be "constructed" out of the initial real elements and their relations; and he proposed to designate these configurations as "imaginary elements," since the relations between these complex configurations are formally identical with the relations between the imaginary elements heretofore regarded as "existing" in their own right. However, instead of *postulating* the existence of these mysterious and paradoxical entities, Von Staudt simply *defined* or *constructed* complex configurations *within* the original domain of real points with the desired geometrical properties. As a consequence, we are led to see that we may reason upon complex configurations of points in exactly the same way as we did upon the points which had been supposed to constitute the subject-matter of geometry. If we ask *why* this is so or how it is possible, we find the answer in the fact that certain *formal* relations or *structures* are common to *points* on the one hand, and *complex configurations of points* on the other.[56]

In exploring this formal structure, therefore, it is immaterial whether we regard its elements as points, as certain configurations of points, or indeed as anything whatsoever. For the valid derivation of theorems depends entirely upon the formal relations supposed to hold, and not upon any intuitive interpretation we happen to give to the terms of the formulations which are the starting point of demonstrative geometry. Von Staudt's method of construction for enlarging the calculus of geometry thus supplements and reinforces the method of implicit definition. Each method leads to the recognition of the fact that in demonstrative geometry the formal pattern of the formulations alone is relevant and that it is not the study of the properties of extension which is the task of this discipline.

3

That pure geometry is neither a science of quantity nor the science of extension is the conclusion which naturally emerges from the development of geometrical ideas thus far traced. Everything in this history points to the view that the only identifiable subject matter which can be assigned to demonstrative geometry is the interconnections of the symbolic operations whose properties have been formally specified. However, the clear emergence of such a doctrine was hindered for a number of years for a number of reasons. Not the least important of these reasons was the common belief, supported by the traditional formulation of geometry and reinforced by the analytic methods of Descartes, that its subject matter was constituted out of inherently simple entities designated as "points." The method of construction as illustrated by Von Staudt's successful interpretation of "imaginary point" as a short-hand designation for complex geometrical relations has an obvious bearing on this dogma; for it followed from Von Staudt's analysis that these relations could be regarded as the primary material for geometrical study with as much reason as the allegedly simple points. Nevertheless, Von Staudt's constructions still assumed "points" as the fundamental elements to describe the complex relations which were then considered as units. The liberation of geometrical terms from their usual but narrow interpretation first required a thoroughgoing denial of the need for absolute simples as the foundation for a demonstrative geometry. Such a liberation was in large measure the consequence of the discovery of the principle of duality and of the manfifold extensions and applications which were made of it.

1) Gergonne. It is difficult to say when it was that mathematicians first noticed that many theorems in geometry can be correlated by pairs on the ground of certain analogies between the relations which they each formulated. As early as the seventeenth century Snellius and Vieta had observed curious symmetries between a number of theorems on spherical triangles.[57] More than a hundred years later Brianchon deduced by projective methods his famous theorem on hexagons circumscribed around conics, and so established a correlate to Pascal's inscribed hexagon theorem. The principle of duality seems almost to leap at us when we reflect upon these two theorems, and it is somewhat difficult to imagine that it escaped the attention of so many excellent men. But Brianchon does not seem to have recognized or explicitly used such a principle, and neither did other members of the school of Monge.[58]

The first mathematician to formulate a general principle of duality for points and lines in a plane and for points and planes in space was undoubtedly Gergonne. In the first of the three major essays he devoted to this subject, he pointed out some important corollaries not noticed by Legendre in the latter's deductions from Euler's theorem on the relation between the number of faces, vertices, and edges in a polyhedron.[59] Gergonne then went on to declare that in general "there is no theorem of this type for which there does not correspond another, deduced from the former by simply interchanging the words "faces" and "vertices."[60] For example, the following two theorems are related to one another in the indicated manner: in every polyhedron the number of *faces* which have an odd number of edges is always even; in every polyhedron the number of *vertices* which have an odd number of edges is always even. In order to emphasize the correspondence between pairs of such theorems which he listed, he printed them in parallel columns in his journal.[61]

A more general explicit formulation of the principle of duality was reserved for the second essay on the subject, published in the following year.

> An extremely striking feature of the geometry which does not depend in any way upon metrical relations [he declared], is that with the exception of some theorems which are themselves symmetrical, for example Euler's theorem on polyhedrons, all the theorems are dual. That is to say, to each theorem in plane geometry there necessarily corresponds another, deduced from it by simply interchanging the two words "points" and "lines"; while in solid geometry the words "points" and "planes" must be interchanged in order to deduce the correlative from a given theorem.[62]

Gergonne complained that even in recent years geometers had neglected to notice this fact: "with so little philosophy is the study of the science pursued even in our day"; and he hoped that his writing a special essay on the subject would contribute to remove this neglect. He pointed out that only half of the theorems he developed in this essay need be demonstrated *directly*, because the other half can be obtained from the former by means of the principle he enunciated. However, he supplied direct proofs for *all* the theorems, in order tho show "that there exists the same correspondence between the *proofs* of two dual theorems as between the *theorems* themselves."[63] Indeed, he even went further: he stated in parallel columns two sets of propositions assumed as axiomatic, and followed these with a large number of theorems together with their proofs. He employed no diagrams because he maintained that "what is relevant here is the logical deduction"; and he believed that he had succeeded in placing beyond every doubt the important point in the philosophy of mathematics "that there is an important part of geometry in which the theorems, as well as the proofs for them, correspond to one another exactly, two by two, and that this is so *because of the very nature of extension itself.*"[64]

So much for the *statement* of the principle of duality. What, however, were the precise *grounds* advanced for it by Gergonne? Was it simply a happy guess, an inductive leap; was it an *a priori* grasp of a fundamental truth of extension; or was the principle demonstrable with the help of the premises of projective geometry? No unambiguous answer to this question can be found in Gergonne, and his views shifted from one alternative to another. Nevertheless, what he had to say on the matter is both interesting and instructive.

In his first essay on the principle of duality he definitely regarded the part of projective geometry known as the theory of reciprocal polars as the mainstay of his principle.[65] But in terms of such a defense, the principle can be taken to formulate only a very special feature present in a very special branch of geometry as a consequence of a very special mode of correspondence which had been established between lines and points by the theory of reciprocal polars. However, such a characterization of the principle was not congenial to Gergonne, for he seems to have thought of it in a more general way and sought to find a basis for it in the "nature of extension." For example, he had occasion to review, under a pseudonym, his second essay on the principle of duality, and he explained how points and lines may be made to correspond in a plane with the help of the theory of polars. He declared, however, that "if the theory of poles, polars, and polar planes makes evident this duality within an important branch of geometry, this duality most assuredly does not hold in virtue of this theory, but holds because of the very nature of extension."[66] Hence the theory of polars may lead us to recognize the principle of duality, but does not constitute its meaning nor the grounds of its validity.

On the other hand, how that knowledge is achieved of the intrinsic character of extension which the principle expresses, if not with the help of the theory of polars, Gergonne does not make clear. He gives the impression that this knowledge is a consequence of some form of *a priori* insight into extension, and so manages to supply rather than resolve a mystery. His most satisfactory defense of the principle, independent of considerations drawn from the theory of polars, is to be found in his second essay on the subject, In that version, as already mentioned, he presented in parallel columns two sets of axioms, the members of the two sets being duals of one another; consequently, the two sets of axioms can be regarded as generating two systems of propositions, so that in one system points are the fundamental elements (lines and planes being defined in their terms), while in the other system planes are the fundamental elements (points and lines being defined in their terms). Now since for every statement in the proof of any theorem in one system there corresponds a dual statement in the proof of the dual of that theorem in the second system, Gergonne was entitled to claim that the principle of duality did not rest upon the theory of polars. It may therefore be taken as expressing the manifest structural identity of the two orders of logical deduction: indeed, if the *terms* "point," "line," "plane" are freed from their ordinary references and are employed simply as *variables* with no specific "meanings," the two systems exhibit an identical pattern of interrelations of elements.

Such an approach to the principle of duality approximates very closely to modern methods of establishing the principle.[67] Had he pursued this line of approach, Gergonne would have easily arrived at the conception, familiar to him in the germ, that "axioms" in pure geometry are not statements which are either true or false, but "implicit definitions" capable of having different concrete meanings associated with the variable-terms they contain. In a word, he would have recognized that the principle of duality does not hold because of the nature of *extension*, and that it formulates the fact of structural identity in systems of isomorphically related elements. In fact, however, as is evident from the above, his views fluctuated through a whole gamut of alternatives: the principle was for him an organizing device for discovering new theorems and an instrument for dispensing with direct proofs for half the theorems in projective geometry;[68] he thought of it as expressing an absolute property of extension and as capable of an *a priori* justification; he thought of it as a simple consequence of the theory of polars; and he also thought of it as a consequence of the fact that two systems of statements had common logical properties, irrespective of the specific, concrete meanings of the statements.

It is because Gergonne was at least conscious of the larger bearings of the principle of duality that he is entitled to the claim of being the discoverer of the principle. Poncelet also made that claim. For he declared that his theory of reciprocal polars enabled him to correlate every projective theorem with another one as general as the first, so that the whole business of discovering new theorems, when certain other ones had already been established, "is reduced to a sort of mechanism" by substituting one set of words or letters for another set of words and letters.[69] Since the theory of polars is essentially a method for establishing a one-to-one correspondence between points and lines and conversely, Poncelet claimed that the obvious duality of theorems was his discovery.[70] Gergonne replied that he had pointed out the facts of duality even before the appearance of Poncelet's publications, and that the latter had not adequately stressed these matters even subsequently. He also declared that the principle of duality can suffer no exceptions because it has "an *absolute generality,*"[71] and made a vigorous plea for a suitable language in which dual properties could be clearly expressed: for "when the language of a science is well-constructed, logical deductions are made with such ease that the human mind advances to new truths, by itself, so to speak."[72]

There is no doubt that Poncelet did discover a special device for correlating geometrical figures which enabled him to establish the reciprocal interrelations of projective theorems. But it is also clear, even from the above brief sketch of a long and heated controversy, that Poncelet did not even glimpse the full logical import of the principle of duality. He explicitly rejected the principle as an autonomous principle which was independent of the theory of polars, on the understandable ground that this theory alone enabled anyone to demonstrate rigorously special cases of dual relationships. "Duality" is a word, he remarked, "which according to me has no sense in the philosophy of mathematics, when it is intended to be used as a type of general principle independent of the theory which alone justifies it."[73] One can sympathize with this dictum whatever one's opinions are on the matter of priority between Gergonne and Poncelet, for it formulates a demand for rational proof and is directed against obscurantism in every form. Nevertheless, the genuine issue between Gergonne and Poncelet was not the question of priority; that issue was whether the principle of duality can be "justified" only on the ground of the theory of polars. Poncelet was of the opinion that it could be justified in no other way; Gergonne thought that it could, although the grounds he advanced for it as a general principle not depending on the theory of polars are far from clear. But even though Gergonne may not have offered a clear and sound foundation for his principle, even though his view that the principle expressed a natural law of extension is inadequate, he was quite correct in maintaining that it expressed a more general fact of logical interconnection than Poncelet was willing to allow.[74] And there is no doubt that on this matter Gergonne was right and Poncelet was wrong.

Steiner, whose contributions to the subject are of first-rate importance, evaluated the issue between them in so far as it bore directly on pure synthetic geometry. As he declared in the preface of his own book,

> I believe that the controversy over the priority of the principle of duality and the theory of reciprocal polars in unambiguously resolved in the present work . . . Dual relations in geometry appear simultaneously with the introduction of the fundamental elements, while the theory of reciprocal polars makes its appearance only subsequently, as a consequence of certain relations between the fundamental elements. Gergonne's principle thus lies nearer to the essence of things, and in this respect is prior to the theory of polars.[75]

2. Chasles. It is of some interest to examine the views of Chasles, another French contemporary of Gergonne, on the relation of the principle of duality to Poncelet's theory of reciprocal polars.

It was clear to Chasles at the outset that the theory of polars is only *one* method among many others for establishing correspondences between figures and so making evident dual properties.[76] After enumerating various other "transformations" (i.e., modes of correlation) which exhibit dual theorems, he concluded that "the fundamental reason for this property of duality inhering in the forms of extension" must be sought in a general theory of the transformation of figures. What then is the general character of the transformations which reveal dualities, "independently of every special doctrine"? Chasles found that what is relevant in the theory of polar transformations for establishing dual theorems is the following: in a space of three dimensions a plane is made to correspond to a point and conversely; and the correspondence is such that planes, which in the second of two correlated figures correspond to *coplanar* points in the first figure, are themselves *copunctual.*

It is this latter condition which Chasles regarded as particularly important, for it distinguishes one type of transformations from an infinity of others which also establish a correspondence between points and planes. But it so happens that while the transformations of the theory of polars satisfy this latter condition, they have according to Chasles further properties which are really quite "accidental" to the fact that dual relationships are established by them.[77] Since Chasles could cite transformations which satisfied the second part of the above stated condition but which did not have those further "accidental properties," he concluded that the theory of reciprocal polars does not supply the most general type of dual transformations. The most general transformations in the sense that Chasles thought of them are the "linear" or projective ones;[78] and he pointed out that these transformations entail dual theorems not only in geometry but in pure algebra as well.[79] The obvious consequence of this analysis is that Gergonne's belief in the principle of duality as based upon the alleged nature of extension is certainly mistaken, since the principle could be shown to follow from considerations which did not apply exclusively to space.

Chasles was quick to see the far-reaching consequences of his conclusions. He pointed out that two kinds of geometries must be distinguished.

> In the first, the *point* is the unit or element, the *monad* out of which are constructed the other parts of extension; it is the basis of the philosophy of ancient and analytic geometry. In the second geometry, it is the *line* or *plane* (according as one is working in the plane or in space) which is the *primitive element* or unit out of which other parts of extension are formed. This classification . . . based on two primitive ideas which are essentially different, . . . seems to me of greatest importance for geometry. But we may extend it to many other parts of mathematics, and, if we bear in mind the beautiful law of duality of spatial figures and are guided by the dualism of the primitive element which is taken as the point of departure in geometry, we shall be led to look for similar states of affairs in them. We may find an illustration of such a duality in a new analytic geometry, analogous to that of Descartes, in which the *plane* plays the same role as the *point* does in his. The same ideas of duality may also be applied to mechanics. The primitive element of bodies to which the first principles of this science are applied, is the mathematical *point*. Now are we not entitled to suppose that by taking the *plane* as the element of extension instead of the point, we shall be led to other doctrines, thus obtaining, so to speak, a new science? And if there should exist a unique principle for passing from the newer science to the older one, . . . this principle would become the foundation for duality in the science of motion. . . . One can thus find in the different branches of mathematics other laws of duality, founded on other principles; and one is thus led to admit . . . that a *universal dualism* is the great law of nature and dominates all parts of human knowledge.[80]

Chasles was no doubt intoxicated by a vision of the operation of a "universal dualism" in all things. But he was sober enough to see clearly the essential features and the import of what he saw. He understood that the age-old conception of points as the necessary primitive elements of geometry was a limitation and a prejudice, and that other elements could be taken as basic with equal right. He also recognized that by means of suitable "transformations," a theory which apparently concerned itself *exclusively* with certain specified phases of a subject matter, could be reinterpreted so as to yield a theoretical account of the interrelations of other phases: this was, indeed, the substance of his vision. He was therefore prepared to admit that the definition of geometry as the science of the measurement of extension must be reconsidered; it is not exclusively the science of *magnitude*, since it is also, and perhaps in a preeminent sense, the science of *order*.[81] He did not quite grasp the point, as did Grassmann, that a pure geometry is the science of "forms," and that like any other branch of pure mathematics has for its subject matter symbolically represented "possible transformations." Nor did he supply the detailed techniques for the construction of systems of geometry which assumed as basic other elements than points: this was reserved for other contemporaries and successors. However, in pointing out the fact that the principle of duality rests on the general features of correspondences of transformations of various elements, he advanced the logical analysis of geometry beyond anything achieved by Gergonne or Poncelet. His discussion of the principle of duality prepared the ground for the unambiguous recognition of the fact that pure geometry is a science of order whose theorems formulate structural identities of any sets of isomorphic relations.

3. Plücker. The techniques of realizing the geometries which take elements other than points as basic, were being supplied contemporaneously with Chasles' work. A more complete generalization of the principle of duality and a thorough exploitation of the suggestions he made were carried out, independently of Chasles, by Moebius and Plücker. It will be sufficient to consider explicitly, though briefly, only the work of the latter.[82]

Plücker's approach to geometry was from the side of algebraic analysis, and the equivalent of the principle of duality was first suggested to him by the symmetry with which the letters "x," "y," and "u," "v," respectively, entered into the equation

$$ux + vy + 1 = 0.$$

On the usual interpretation which is assigned to equations of this form in analytic geometry, "u" and "v" are *constants* determining a class of *points* (with variable coordinates "x" and "y") which lie on the straight line whose intercepts with the coordinate axes are the negative reciprocals of u and v. But Plücker noted that if "x" and "y" are taken as fixed and "u" and "v" as variable, the equation determines a class of *lines* (with variable coordinates "u" and "v") which pass through the point whose coordinates are x and y. On the first interpretation the equation specifies a *line* as a locus of *points;* on the second interpretation it specifies a *point* as a locus of *lines*. Consequently, instead of assuming points as the basic elements and lines as derivative from them, as is done in the traditional analytical geometry, Plücker worked out the alternative method suggested by these observations: he assumed lines as basic and points as secondary. As a consequence, points and lines may each be regarded with equal right as "fundamental" elements for a plane geometry; in three-dimensional space points and planes occupy the same sort of equal rank. Since, therefore, the two sets of parameters "u," "v" and "x," "y" enter symmetrically into the bilinear equation

$$ux + vy + 1 = 0$$

for plane geometry (with analogous symmetry for the equation

$$ux + vy + wz + 1 = 0$$

for space), so that it may be regarded in the two-fold manner just indicated, every projective theorem about points can be immediately matched by an analogous theorem about lines and conversely; the principle of duality is simply a consequence of this symmetry.[83]

However, Plücker was not content with this restricted form of the principle of duality; by applying similar considerations as the above, but relevant to equations of degree greater than one, he established his general "principle of reciprocity" independently of the work of the French school. Indeed, this principle was much more powerful than anything discovered by the latter, and was grounded upon very different kinds of considerations. Thus, Plücker observed that the principle of duality as employed by Gergonne and Poncelet was tied down to the theory of reciprocal polars, a restriction which he thought was clearly unneccessary in the light of his own analysis: for the dualities in question were a consequence of the formal relations between sets of coordinates, and were not, as Gergonne believed, an absolute property of extension—a belief which Plücker characterized as a "metaphysical abstraction."[84] While an adequate account of Plücker's generalization of the principle of duality cannot be given in the present essay, some inkling of his method and results may be obtained from the following sketch. In the plane, the equation

$$ax^2 + bxy + cy^2 + dx + ey + 1 = 0$$

is usually taken to represent a conic, where "x" and "y" are variable point-coordinates; points are here taken as basic, so that the conic is a locus of points. But just as points may be regarded as the loci of lines when lines are assumed as the fundamental elements of the system, so Plücker proposed to take *conics* as the primitive configurations of a new system; and just as a line is completely specified by two line-coordinates "u," "v," so a conic may be completely specified by the five conical-coordinates "a," "b," "c," "d," and "e." Consequently, every equation between variable conical coordinates will represent a configuration in the plane which is a locus of conic sections; that is to say, the configuration will be specified in terms of relations between conics, rather than in terms of relations between points or even straight lines.

An immediate consequence of this altered point of view is that the dimensionality of a "space," e.g., of the plane, depends upon the elements with respect to which dimensionality is predicated: thus, the plane is two-dimensional with respect to points and also with respect to lines; but it is five-dimensional with respect to conics, since five independent coordinates are required to specify completely a conic section. In general, the dimensionality of a manifold is relative to the choice of configuration as element. Plücker himself worked out a part of the goemetry of "ordinary three-dimensional space" with the line as the fundamental type of configuration; consequently, since four independent coordinates are required to determine a line in this space, he in fact developed a "four-dimensional" geometry. But perhaps an even more remarkable consequence of abandoning points as the necessary primary elements was the extension of the principle of duality or reciprocity to make it exhibit dualities not only between points and lines in the plane (or points and planes in space), but also between *any* pairs of configurations whose dimension-numbers (defined as the number of independent coordinates required to specify the configuration) with respect to the manifolds in which they occur is the same. As Plücker remarked, "it follows from the principle of reciprocity that any projective theorem about lines may be carried over directly to all such algebraic or transcendental curves whose equations, referred to any coordinate systems, are linear with respect to two of its constants, so that the theorem can be multiplied indefinitely."[85] The essential factor in all this is the form of the equations with which one is working. For

> every geometrical relation is to be viewed as the pictorial representation of an analytic relation, which, irrespective of every interpretation, has its independent validity. Consequently, the principle of reciprocity properly belongs to analysis, and only because we are accustomed . . . to express the matter in geometrical language, does it seem to be an exclusively geometrical principle.[86]

The failure to understand this was Gergonne's error: the roots of the principle of duality are not to be found in any absolute property of extension.

> If we carry through the proof of a theorem concerning straight lines (using the letters "*a*," "*b*," "*c*," . . . to designate linear forms in two variables for representing such lines), we have in fact demonstrated an untold number of theorems. For if by the letters "*a*," "*b*," "*c*," . . . we no longer designate linear expressions but any general functions in two variables, provided they are of the same degree, the conditional equations
>
> $$F(a, b, \ldots u, v, \ldots) = 0$$
>
> [which formulate the relations which hold between straight lines on the initial hypothesis] as well as all the equations derived from them retain their meaning. . . . If we have such a proof-schema (*Beweisschema*) we may relate it to lines of any arbitrary order. . . . We may therefore carry over every theorem in projective line geometry to curves of any arbitrary degree.[87]

In brief, the proof of one theorem is transformed into the proof of another when the symbols occurring in the analytic proof-schema are interpreted differently:[88] for "when an analytic proof of a theorem is once given, as many different theorems are proved at the same time as there are different systems of coordinates."[89]

The central logical idea which thus emerges from Plücker's work is that a mathematical proof depends solely on the stipulated connections between "abstract" or variable signs, and that the interpretations which may be given to those signs do not affect the validity of a demonstration. Plücker's great technical achievement was to have provided a method for translating statements about one set of geometrical configurations into other statements about "intuitively" or pictorially *different* configurations. That technique rested upon his discovery that any configuration may play the role of a "fundamental" or basic element in the construction of a pure geometry. That is why the abandonment of the conception of points as inherent simples was such an important step in the development of modern ideas in logic and mathematics. To it we owe in part the conception that pure geometry is a *formal* discipline operating with variables, which *may* but *need not* be interpreted, and that pure geometry, in thus becoming abstract and without an intuitive content, is nevertheless an instrument for formulating identical structures in intuitively different subject matters.[90]

4

Two distinct strands in the development of modern geometry have now been partially unraveled which lead directly to a radical overhauling of traditional philosophies of mathematics. It has been shown how Poncelet's principle of continuity became a heuristic device for constructing new symbolic systems employing the language of geometry, and how, in order to establish the material thus obtained on a sound logical basis, the methods of implicit definitions and construction had been developed to serve as powerful techniques for further reconstructions and extensions of the traditional system of pure geometry. It has also been shown how the discovery of impressive dualities between the familiar theorems of geometry led to a revision of the belief in inherent and absolute simples, and how as a consequence pure geometry was gradually recognized as a formal discipline whose identifiable subject matter consists of the interrelations of symbolic operations and symbolic relations.

These parallel developments obviously support each other and work themselves out in such a way that a common philosophy of pure mathematics is the end-product. On the one hand, however, this common philosophy was not explicitly formulated by the writers examined thus far, so that something should be said concerning the ideas of later students who have built on the foundations laid down by those who have already been considered. On the other hand, the problems which have been discussed pertain primarily to formal or demonstrative geometry, and something should also be said about the emergence of altered views concerning the relation of formal geometry to the subject matter traditionally alleged to belong to it. In the present section a brief survey will be made of the writings of such relatively recent students as Pasch, Klein, and Poincaré, in order to supply a few of these omissions.

1) The writings of Pasch contain one of the first clear expressions of the view that pure geometry is a "hypothetico-deductive" system, whose "axoims" are in effect simply "implicit definitions" of the terms they contain. Pasch illustrates in his own person the possibility of a mathematician being a thoroughgoing empiricist (in fact, an "old-fashioned" empiricist) with reference to the source and validity of geometry as a *natural science,* while maintaining *pure geometry* to be an autonomous discipline which can and must be cultivated without appeal to matters of empirical fact. He denied that geometry as a natural science required or could have an *a priori* foundation, and maintained that the evidence for the truth of its axioms rested on the facts of sensory intuition; but he also claimed that the validity of the deductions of theorems is completely independent of the intuitive meanings of the terms which they contain.

Pasch's empirical standpoint is clearly indicated in the prefatory remarks to his important book on geometry. "Geometrical concepts," he there says, "are a special group of those concepts which serve to describe the external world; they refer to the form, magnitude, and mutual positions of bodies. By introducing numerical concepts, relations between geometrical concepts become manifest so that they can be recognized by observation. The standpoint is thus indicated which is assumed in the following, according to which geometry is a part of natural science." Pasch then goes on to explain what is to be understood by "point," "line," and "plane" when these are taken as *descriptive* terms with an empirical denotation; for example, *a physical body* is to be called a "point" if its subdivision into parts is incompatible with the limits set by actual observation. He is careful to add, however, that "the application of these ideas is associated with a certain amount of uncertainty, as is the case with almost all concepts which we have constructed for conceiving the external world."[91]

A relatively small number of such ideas (which, strictly speaking, are *undefined,* although their *use* is described) can be used to define all the others which occur in geometry, and a relatively small number of propositions, themselves not demonstrated, can serve to demonstrate all other propositions: they are what Pasch calls the "nuclear" concepts and propositions, respectively. Nuclear propositions are based on observations which have been indefinitely repeated, and which have been established so securely that mankind has forgotten their origin. But although the various concepts and the relations between them are grounded in observation, the deduction of theorems does not depend upon this fact.

> When Euclid states in his *Elements* "A point is what has no parts; a line is length without breadth; a straight line is that which lies evenly with the points on itself," he is not explaining these concepts through properties of which any use can be made, and which in fact are not employed by him in the subsequent development. Indeed, at no point does the argument rest upon these declarations, by which the reader, who can learn nothing about the nuclear concepts from the *Elements* if his imagination has not been previously trained by repeated observations, may at best be reminded of the appropriate intuitions, so that he may limit or augment them in accordance with the demands of science.

Although the nuclear propositions are intended to formulate the facts of experience and to state completely the empirical material which mathematics is required to analyze and elaborate, once the propositions have been stated and accepted there is no further need to refer back to sensory experience. A theorem is demonstrated only then when the proof is completely independent of every diagram.[92]

These commitments place Pasch on familiar ground hallowed by tradition. But as he advances in the construction of his system of geometry, the system becomes increasingly formal and abstract so that in the end the nuclear concepts and propositions are robbed of all concrete reference. In the first place, the nuclear terms "point," "line," etc., which originally denoted bodies given to sensory intuition, are given new meanings and wrenched from their anchorage in the materials of the initial intuitions. The word "point" in its extended sense, for example, is taken to refer to a bundle of lines, that is to the set of all rays emerging from a common center.

> The identical procedure is repeatedly found in mathematics on similar occasions. Thus, in the process of gradually extending the concept associated with the word "number," one is compelled to employ the expression "real positive whole number," although the word "number" without

> qualification was sufficient. . . . But just as it is only for real numbers that one of them is designated as "greater" than another, so it is only for three proper points lying in a straight line that one of them can be said to lie "between the others." This latter concept and all further ones defined by its means are restricted to proper points. Nevertheless, the definitions and relations already laid down (for proper points) retain their validity (for the extended use of the word "point").[93]

Similar extensions are given by Pasch to the meanings of the words "line," "plane," and even "between," with the consequence that while the nuclear propositions are valid for proper points, lines, and planes only under certain restricting conditions, these propositions become valid without such qualifications when the terms are assigned these extended meanings. But as Pasch clearly states, "the attempt to bring the maximum number of configurations within a common point of view, so that they may all be considered in a uniform way, has progressively forced the 'proper' meanings of the elementary geometrical terms into the background."[94] The historical development, beginning with the introduction of improper elements by Poncelet and culminating in Von Staudt's interpretation of such elements with the help of the method of construction, is thus repeated in Pasch's book, though consciously and with a full understanding of the logic involved.

In the second place, the principle of duality for projective properties, which emerges in the course of the discussion, is taken by Pasch to indicate that the procedures of pure geometry are independent of the meanings which may accidentally be attached to the nuclear "concepts." The proofs of two dual theorems differ in no essential way, so that the validity of what is in fact an identical pattern of proof cannot be affected by the meanings of the terms involved.[95]

> Indeed [Pasch declares], if geometry is to be really deductive, the deduction must everywhere be independent of the *meaning* of geometrical concepts, just as it must be independent of the diagrams; only the *relations* specified in the propositions and definitions employed may legitimately be taken into account. During the deduction it is useful and legitimate, but in *no* way necessary, to think of the meanings of the terms; in fact, if it is necessary to do so, the inadequacy of the proof is made manifest. If, however, a theorem is rigorously derived from a set of propositions—the *basic* set—the deduction has a value which goes beyond its original purpose. For if, on replacing the geometric terms in the basic set of propositions by certain other terms, true propositions are obtained, then corresponding re-

placements may be made in the theorem; in this way we obtain new theorems as consequences of the altered basic propositions without having to repeat the proof.[96]

It is therefore important, in Pasch's opinion, that pure geometry be *formal* in the strict sense. Everyone is aware, he pointed out, of the dangers which threaten the geometer who uses diagrams and other sensory images; but few seem to be equally aware of the traps which we lay for ourselves when we employ common words to designate mathematical concepts. For such words have many associated meanings not relevant to the task of a rigorously deductive science, and these associated meanings sway us to the detriment of rigor. To avoid these handicaps it is therefore desirable to *formalize* the set of nuclear propositions; that is, we ought to replace them by a series of expressions in which the "geometrical concepts" of the propositions have been replaced by arbitrarily selected marks, whose sole function is to serve as "places" or blanks to be filled in as occasion may warrant. The result of such a formalization is an "empty frame," which expresses the structure of the set of nuclear propositions and which is alone relevant to the task of pure geometry.[97] "A statement R^1 is a consequence of *B* only because its derivation from the latter is completely independent of the meanings of the geometrical concepts occurring in it, so that the proof can be carried through without support from an actual or imagined diagram or from any sort of 'intuition.' A proof which does not meet this condition is no mathematical proof."[98]

However, Pasch did not wish to deny that pure geometry may be cultivated even when terms like "point" appear in it as undefined or unexplained. In that context, however, these familiar terms do not denote anything "real" or "actual"; and when we think of pure geometry from the point of view of a natural science, the "terms" and "propositions" of pure geometry have no identifiable subject matter, and may at best be taken as "hypothetical concepts" and "hypothetical propositions." In order that the pure, formalized system be "applied" to natural objects, so that the axioms become definite statements about the latter, the hitherto unspecified terms must be "interpreted": the "interpretations" being simply correlations between such an expression as "point" and appropriate kinds of bodies.[99]

Pasch's interest in the philosophy of geometry was thus divided between two problems: the character of geometry as a demonstrative science; and the status of geometry as a natural science. His recognition of the distinction between pure and applied mathematics which is implied by his formulation of issues was by itself a contribution of great importance, and the turn which his answers took was almost an inevitable consequence of that distinction. On the issue of geometry as a demonstrative science he was decidedly a formalist, maintaining that the rules of formal logic and these rules alone were sufficient for developing the discipline and evaluating its findings. He was perforce compelled to break with those traditions according to which intuitions of some kind, whether sensory or "pure," are required for geometry, or which regard space, or extension, or the magnitudes of bodies as its subject matter. On the second issue he was an empiricist and indeed an empiricist of the following sort: he gave a *genetic* account for the "axioms" with which mathematicians were as a matter of fact concerned, and believed they were suggested by the materials of sensory intuition; he declared that the axioms of applied geometry formulated relations between *bodies;* and he claimed that the factual truth of these axioms could be and is established by *sensory observation* on bodies. In this connection, also, Pasch was at odds with traditions both philosophical and mathematical: applied geometry was a natural science not of space or extension as such, but of bodies. In spite of the clarity with which Pasch stated his views, however, his empiricism was essentially naive; and he seems never to have faced the difficulties of establishing the axioms of applied geometry as factual truths on the basis of an empiricism like that of Mill. But this was a matter upon which sufficient clarity was not obtained until a number of years had elapsed after Pasch first wrote.

2) Pasch's book was the first of a series of works on geometry which formulated pure geometry as a system of logical relations between variables. The traits of logical purity and rigor which his book illustrated became the standard for all subsequent essays in geometry. No work thereafter held the attention of students of the subject which did not begin with a careful enumeration of undefined or primitive terms and unproved or primitive statements; and which did not satisfy the condition that all further terms be defined; and all further statements proved, solely by means of this primitive base. Indeed, if anything, later writers have set for themselves standards of an even greater rigor of formalization than Pasch's work exhibits.

In Italy, Pasch's treatise became known through Peano's translation of the substance of the *Neuere Geometrie* into the intuitively opaque notation of mathematical logic. No phase of Pasch's empiricism is reflected in Peano's version, and pure geometry in his hands became a calculus operating upon variables formally stipulated to be related to one another in certain ways. The pattern of formal relations which is alone relevant to the validity of demonstrations thus receives a systematic expression, in even more forceful one than Pasch managed to give it.[100] A flourishing school developed around Peano which aimed to systematize in like manner not only geometry but all branches of pure mathematics; and Pieri, a member of this group, characterized neatly its conception of geometry by designating it as a "hypothetico-deductive" system.[101] However, not all Italian mathematicians subscribed to the thoroughgoing formalism which Peano initiated, and indeed most of the great Italian geometers of the last and present centuries have consistently sought to find an intuitive interpretation for the analytic formulas of other continental mathematicians.

Veronese is a case in point. He was not sympathetic to the somewhat crude empiricism of Pasch. But he believed that the n-dimensional geometry he developed had its roots in our spatial intuitions; he regarded geometry to be the most exact experimental science, refused to "reduce" mathematics to a system of conventions for the manipulations of signs, and complained that the analytic method in geometry, as distinct from the synthetic, did not make explicit the specifically "geometrical" relations which he thought were involved in the proofs of theorems.[102] Nevertheless, he did not escape the influence of Pasch's logical demands; and his book contains a remarkably careful, and in places exaggerated, embodiment of the formal conditions for a deductive system.[103] He pointed out that the raw material of sense must be elaborated intellectually, and that in order to effect the widest range of hypotheses for understanding the objective order of things, we must be ready to consider "ideal forms" which may not be apprehended by sense alone. He admitted that the construction of geometries whose dimensionality is greater than three involves a process in which "intuition is fused with pure abstraction," since the "configurations" in such "spaces" cannot be completely intuited.[104] In point of fact, the "geometry" of the hyper-spaces which Veronese obtained was modeled upon the familiar three-dimensional system, so that the latter was a "suggestive science" for the former in the sense of Peacock and Hankel; Veronese's n-dimensional geometry" is a calculus containing the familiar terms of familiar geometry which relates them in ways formally analogous to types of connections present in the latter. The familiar terms are not assigned an identifiable denotation, and are therefore simply implicitly defined by the relations into which they enter: they function as variables in a matrix constructed with the glue of logical constants. Just such a view is sponsored by Enriques, whose examination of the principles of geometry terminates in this conclusion.[105]

However, the work which perhaps more than all others has been the basis for contemporary views on pure geometry is Hilbert's *Grundlagen der Geometrie.* Although the motto for its brief introduction is taken from Kant—"All human knowledge begins with intuitions, proceeds to concepts, and terminates in ideas"—its effect has been to fortify a decidedly anti-Kantian philosophy of mathematics. For while the explicitly stated objective of the book is to supply a careful logical analysis of our spatial intuitions, in the details of its execution all reference to intuition is rigorously excluded: the only indication which may be found in it that Hilbert is ostensibly concerned with the traditional subject matter of geometry is his use of traditional terminology.

Hilbert begins his system with the following explanations:

> We consider three different systems of things: the things of the *first* system we call *points;* the things of the *second* system we call *lines;* the things of the *third* system we call *planes.* . . . We consider points, lines, and planes in certain mutual relations, and designate these relations by words like "lie," "between," "congruent," "parallel," "continuous"; the exact and mathematically adequate description of these relations is given by the *axioms of geometry.*"[106]

It is evident, therefore, that the elements of the system are specified in no way other than by requiring that they satisfy the mutual relations assumed to hold between them, while these relations are themselves not further specified. Neither the terms for the three kinds of things nor the words for the different relations are given any intuitive content officially. The "axioms" simply state the formal properties of these "relations" by specifying the different combinations allowed for them; and although the *words* Hilbert employs to designate these relations have obvious intuitive associations, nothing would be altered in any portion of the book if he had used other words or signs in their place.

It is also true that Hilbert explicitly declares that each of his five groups of "axioms" expresses certain fundamental facts of our spatial intuitions. Nonetheless, this reference to the alleged source of the axioms is essentially a biographic statement, and in no way controls the sequence of the theorems derived or the validity of the derivations. Indeed, Hilbert himself tacitly recognizes that he is occupied throughout his treatise with term- and relation-variables: for example, the independence of the axioms from one another is established by "interpreting" the various expressions which have been enumerated as designating such "things" as algebraic functions, real numbers, and other "entities" of analysis.[107]

The formalization of geometry so that it can be construed as a hypothetico-deductive system is only one example of the persistent attempt to axiomatize the different branches of mathematics, and to view them, at least for purposes of further analysis, as so many "symbolic" systems with no specific reference or application. This is one of the culminating points of the history of pure geometry. However, it is not only the terminus of a development; it is also the starting point of a new history and of a relatively new systematic discipline, whose aim is to explore the relations between different operations upon symbols as employed in various formalized systems. Into this subject, called by Hilbert "meta-mathematics" and by others "meta-logic" and "syntax," this essay cannot enter. The sort of questions considered by meta-mathematics, e.g., the consistency and independence of axioms, have been discussed in antiquity and have been cultivated ever since by mathematicians in every age; but such problems have been clearly formulated and systematically explored only after pure geometry had been freed from its traditional associations with space, and only after its character as a calculus had been isolated from its applications.

3) While the autonomy of pure geometry became firmly established by the triumph of the axiomatic method, the relation of the formal, symbolic systems (which thus came to be taken as constituting pure mathematics) to the concrete materials of experience was not thereby clarified. Some of this requisite clarification was supplied in the context of discussing the import of so-called non-Euclidean geometries, no small portion of which may be found in the writings of Felix Klein and Henri Poincaré.

The history of non-Euclidean geometry is too well known to bear repetition in this place. The undeniable outcome of the age-old effort to demonstrate Euclid's parallel postulate was the construction of new systems of pure geometry—i.e., calculi employing familiar geometrical language and bearing certain analogies to the calculus of traditional geometry—which, when their terms were interpreted as referring to spatial configurations in the usual way, seemed to be in violent contradiction to the deliverances of habitual "intuitions of space." The work of men like Beltrami, who offered interpretations of these new "geometries" in terms of "intuitions" of familiar surfaces, did much to establish the consistency of these calculi; moreover, it took the wind out of the sails of those who insisted that the new systems could not be construed as "geometries" even though they were consistent calculi, on the ground that they lacked an "intuitive content." However, the precise logical relations to one another of the pure Euclidean, Lobatchevskian, and Riemannian geometries were revealed, not by offering intuitive "models" for each, but by applying considerations drawn from the theory of algebraic invariants (a branch of mathematics much cultivated in England, especially by Cayley and Sylvester) and the mathematical theory of groups. By working out certain technical results obtained by Cayley, Klein succeeded in showing that the differences between the three types of geometry may be viewed as a consequence of the fact that each of these systems employs a different definition of congruence, or measure of distance and angle. A brief outline of Klein's argument is essential for understanding the contributions he made to the subject.

Klein pointed out, in the first place, that projective geometry can be developed consistently without employing any postulate of parallels or axioms of congruence. Numerical coordinates may be introduced, however, by the projective method first developed by Von Staudt, so that pure projective geometry in its analytic version can be viewed as the theory of linear transformations. In the second place, these transformations will leave something invariant: in particular, co-linear points will be transformed into co-linear points, co-punctual lines into co-punctual lines; and the anharmonic or double ratio[108] of four co-linear points, or four co-punctual lines, will remain the same for sets of such elements transformed into each other by projection. But these familiar and elementary facts led Klein to formulate a very powerful generalization of the principle of duality, a generalization which was based upon the intimate connection between a group of transformations and the properties left invariant by them.

The central idea of Klein's famous *Erlangen Programm,* which contained and applied this generalization to many fields, was stated by him as follows: "The geometrical properties characteristic of a geometry remain unchanged by the principal group, and the geometrical properties of a system are characterized by the fact that they remain unchanged by the transformations of the principal group."[109] The sense of this dictum can be briefly explained. The differences between the various geometries are in fact the differences between the relations they explore. For example, the usual geometry studied in school discusses metrical relations, such as the conditions under which line segments, angles, areas, and volumes are equal or unequal; projective geometry studies the conditions under which a set of points remain co-linear or a set of lines remain co-punctal, and so on. Now within each geometry we discover that certain "operations" or transformations may be performed which leave unchanged or invariant the relations which are characteristic of that geometry. Thus, in the ordinary Euclidean geometry, figures may be "translated" or "rotated" without altering the metrical relations established by the theorems, while in projective geometry figures may undergo any series of "projections" without destroying the collinearity of points or altering the anharmonic ratios of concurrent lines. Klein's point, therefore, is that the relations or properties which a geometry explores are those which are invariant under a set of transformations; the invariant properties and permitted transformations mutually determine each other, so that either the invariant properties or the set of transformations may be taken to characterize the geometry. For example, instead of saying that ordinary Euclidean geometry studies relations of magnitude between segments, angles, and so on, we may also say that Euclidean geometry studies those properties which are invariant under translations and rotations.

With the help of this principle Klein was led to the following result. Suppose A is a manifold of elements in which certain properties remain invariant by the group of transformations B; suppose further that the elements of A are made to correspond in any reciprocally unique way with the elements of another manifold A'. It follows that the group of transformations B will be correlated with a group of transformations B' in such a way that the properties left invariant in A by B will correspond to properties left invariant in A' by B'. For example, let A be the manifold of points on a straight line and B the (triply infinite) set of linear transformations which transform the line into itself—that is, which correlate points of A with points of A. Now A can be made to correspond to the points of a conic section A' in a one-to-one manner (by taking a point on the conic as the center of projection and projecting A upon A'), so that the group of transformations B will be made to correspond, also in a one-to-one manner, with the set of linear transformations B' which transform the conic into itself—that is, which correlate points of A' with points of A'. Consequently, for every theorem about the invariant properties of A there will correspond a dual theorem about the invariant properties of A', and conversely; and this simply means that the abstract projective geometry of the points on a straight line is identical with the abstract projective geometry of the points on a conic. But Klein obtained even more astonishing results by recalling Plücker's discovery that the dimensionality of a manifold is relative to the choice of element: that is, a plane is two-dimensional with respect to points but three-dimensional with respect to circles. Klein concluded that when the groups of transformations which characterize respectively two different geometries are abstractly or formally identical, the abstract or formal content of the two systems is also identical, no matter what their respective elements may be. "Every proposition which is obtained for one choice of elements is also a proposition for any other choice."[110] In other words, two geometries, one of which may be about "points" and the other about "circles," are *structurally identical* if their respective groups of transformations are abstractly the same.

But what is the bearing of all this upon the relations between the Euclidean and non-Euclidean geometries? Omitting most of the details, Klein's thesis may be sketched as follows. Consider the group of linear transformations in a plane which leave some arbitrary conic invariant. And given two points a and b, the line determined by them will intersect the fixed conic in two further points. Now form the product of a certain constant k and the logarithm of the anharmonic ratio of these four points, and call it the "distance" between the two points a and b. Klein shows that this function, which is definable entirely in projective terms, satisfies the usual conditions which hold for distance as ordinarily understood; for example, if ab and bc are the distances as defined above between the points *a* and *b* and *c* respectively, then $ab + bc = ca$. Moreover, depending on whether the fixed conic is real, imaginary, or degenerate, the properties which are left invariant by the group of transformations under consideration will be those of Lobatchevskian, Riemannian, and Euclidean pure geometry respectively; and the distance between two points defined as above will coincide with what is ordinarily understood by "distance" in these systems. Analogous projective definitions can be given for the magnitude of angles, areas, and volumes. Klein's general conclusion, therefore, is that the difference between the three types of geometries is entirely *metrical,* and can be viewed as arising from the differences between the definition of such magnitudes as the distance between two points. He pointed out, however, that the *numerical values* of the distance-function for each of the geometries do not differ from one another appreciably in the neighborhood of the origin.[111] Furthermore, since the groups of transformations which leave the distance-functions invariant are abstractly identical, the three types of geometry are also abstractly or structurally identical: for every theorem about an invariant property in one geometry, there is a "dual" theorem about a corresponding invariant property in each of the others. Consequently, the three different metrical geometries considered as abstract calculi do not make contradictory assertions about "space" or "extension," but exhibit in different notations an identical pattern of relations.

In this way Klein definitely settled the formal problems relevant to the interrelations of the three main metrical geometries considered as abstract calculi. Of what significance, however, were these formal studies for the facts of physical measurement and the materials accessible to sensory intuition? This was not a trivial or indifferent matter to Klein. For although he recognized that mathematics may be considered from the point of view of its "immanent" meaning—that is, from the point of view of its being a consistent system—and although his researches on non-Euclidean geometry bore entirely on this question, he believed that mathematics also had a "transient" meaning, and so was relevant to the facts of experience. He confessed that he had "never been able to achieve that indifference to transient problems which characterizes so many contemporary mathematicians interested in axiomatics."[112] Indeed, he seriously doubted the view of Pasch and others that the axioms of geometry could formulate in purely logical fashion the "facts" of spatial relations so completely that thereafter it was no longer necessary to refer back to "spatial intuitions" in the demonstration of theorems; on the contrary, he declared that he was incapable of carrying through any geometrical research on a purely logical basis without recourse to diagrams.[113] It seems, however, that the issue which Klein raised in this connection is in part at least a verbal one; for he did not deny the possibility of a purely algebraic analytic "geometry," though he maintained that such a system is not a "genuine geometry." He did not therefore seriously dispute the claims that demonstrative geometry could be formalized, though he did reveal an interesting biographical fact about his own needs for forming visual models in exploring a formal system employing the language of geometry. On the other hand, what he had to say on the relation of geometrical axioms to the facts of spatial intuition is of great significance, because his views culminated in an extremely radical philosophy of applied geometry.

Klein was frankly empirical in his views on the sources and grounds of validity of geometry, considered as an applied or natural science. In this he was anticipated not only by such philosophical empiricists as Mill (whose influence on mathematics was nil) but also by Helmholtz, whose ideas commanded respect because of his technical achievements in both physics and mathematics. Helmholtz maintained that while the different non-Euclidean geometries presented conceivable theories as to the character of space, nevertheless our daily experience with bodies and our ordinary empirical intuitions of spatial relations established the truth of the Euclidean system.[114] This view seemed much too naive to Klein to be satisfactory. For "spatial intuitions" were regarded by him as "essentially inexact," whether such intuitions were taken as "abstract" apprehensions of the traits of "pure space" or as "concrete" perceptions of physical configurations as in ordinary observation. He maintained, therefore, that an axiom, instead of being a careful formulation of determinate facts of intuition, is a "*postulate* by means of which I *introduce exact statements* into an inexact situation." Consequently, to the extent that the inexactitude of our spatial intuitions is compatible with the assertion of different axioms, any one of these axioms may be postulated; and the choice of one set of axioms rather than another will be, to that extent, arbitrary. "It is just at this point that I regard the non-Euclidean geometries as justified (where by non-Euclidean geometry

I understand the real discipline and not simply the abstract mathematical studies to which it leads). From this point of view it is a matter of course that of equally justifiable systems of axioms we prefer the simplest, and that we operate with Euclidean geometry just for that reason."[115]

This doctrine is repeated by Klein on many occasions and with different emphases.

> Have the axioms their source in experience? Helmholtz has definitely taken this view. But his expositions are inadequate at certain points. It will be readily granted on reflection that experience is a large factor in the origin of the axioms; nevertheless, a point is neglected by Helmholtz which is of particular interest to the mathematician. For a process is involved which is employed by us in exactly the same way in every theoretical treatment of any empirical material, and which will be taken as a matter of course by the natural scientist. I will state my views in general form: *the results of any observation are valid only within certain limits of accuracy and under special conditions; in setting up the axioms we substitute for these results statements having an absolute precision and generality.* In my opinion, the essential nature of the axioms of geometry is to be found in this "idealization" of empirical data. Our postulation of them is limited in its arbitrariness on the one hand only in this, that it mold itself upon the facts of experience, and on the other hand, that no logical contradictions be introduced. Furthermore, what Mach calls the "economy of thought" is a guiding principle which controls the postulation. No one will reasonably hold on to a complicated system of axioms as soon as he recognizes that he can achieve the representation of empirical data with the required accuracy by means of a simpler system. Consequently, to come to the point which is of primary interest for the foundations of geometry: for practical purposes, everyone will assume the formulae of the usual Euclidean geometry and not those, for example, of the Lobatchevskian geometry.[116]

Moreover, Klein points out that we cannot directly know or intuit the properties of "space" in remote or large regions; we are therefore compelled to set up postulates which enable us to deal with them indirectly. However, many of the mathematically formulated "spaces" studied by Lie are topologically different from one another, although all of them are equally compatible with the alleged facts of our experience. If in our theoretical studies of physical space we limit ourselves to the possibilities offered by Euclid, Lobatchevsky, and Riemann and neglect the others, this selection Klein construes as motivated solely by considerations drawn from the principle of economy.[117] "Naive intuition is not exact, while refined intuition is not properly intuition at all, but rises through the logical development from axioms considered as perfectly exact."[118]

In this way Klein laid the basis for a "conventionalistic" view concerning geometry as a natural science. His detailed discussions are not free from obvious unclarities and difficulties connected with the assumption that we are capable of intuiting the properties of "space" as such. Nonetheless, he was one of the first mathematicians to subscribe to a philosophy of geometry which combined two important theses: that geometry may be developed quite formally with attention being paid only to questions of the internal consistency of the system; and that in "applying" such formal systems (that is, in attempting to interpret the variable terms contained in them so that they come to designate experiential facts), alternative systems which are abstractly different present themselves as candidates for such application, between which a choice must be made which is not dictated by the alleged facts. Klein's "empiricism" was different from the sensationalistic variety to the extent at least that for him the axioms of geometry did not simply *report* and formulate immediately apprehended data, but selected and fashioned the materials of experience: axioms were *postulates* for organizing experience and not records of antecedent observations. Morc specifically, the recognition that alternative systems of measurement were possible for organizing our physical experience served as the basis for the explicit conventionalism of Poincaré and prepared the ground for the modern physical theory of relativity.

4) Poincaré's philosophy of geometry brought together into a unified whole the different conclusions reached by the series of researches which have been outlined above. It will be worthwhile to enumerate some of the strands in the argument as developed thus far, before discussing Poincaré's synthesis. In the first place, it has been shown that pure geometry is not primarily a science of quantity or measurement, because geometry frequently explores relations of elements to which considerations of magnitude are not relevant. In the second place, pure geometry is not exclusively a science of extension of space, because elements, relations between elements, and operations upon elements are discussed by it which cannot be identified as belonging to "extension" or "space" when these terms are understood in any ordinary sense. In the third place, pure geometry cannot be regarded as a discipline which has any identifiable subject matter

other than that of its own symbolic operations: as a demonstrative science it engages in the transformation of symbolic expressions and is not concerned with the possible interpretations of such symbols. Pure geometry is thus an autonomous discipline, and while its operations and assumed relations may be suggested by the materials of sensory intuition, it can nonetheless be cultivated in independence of the latter. In the fourth place, the symbolic systems which constitute pure geometry can be interpreted in many different ways, each interpretation offering a qualitatively and intuitively different subject matter as the referent for the abstract system. A system of pure geometry may therefore be viewed as representing a pattern or structure which is identically the same in different subject matters; the various formulations of a principle of duality are so many testimonials for this fact. And in the fifth place, the materials of sensory intuition do not present themselves as logically ordered and as sharply demarcated. They may, therefore, be organized on the basis of alternative modes of arrangement, the choice of the ordering principle being limited only partly by the determinable traits of a subject matter and partly by the desire to achieve the maximum convenience and coherence in our theoretical knowledge. Purely abstract studies themselves suggest different possibilities as to how procedures of measurement may be instituted; each of these possibilities is logically coherent, and the choice of one alternative rather than another cannot be challenged by the immediate deliverances of an alleged intuition of spatial configurations.

Poincaré gathered up these different strands of a discussion begun a century ago to support a radical philosophy of geometry. His general conclusion was that statements such as "*Space* is Euclidean" or "*Space* is Non-Euclidean" are not true or false but strictly *meaningless,* and that the axioms of geometry are not empirical or *a priori* truths but "conventions" or "disguised definitions" for the ordering and measurement of bodies.[119] In this way Poincaré proposed to cut the ground from under the major disputants as to the status of geometry as a natural science: empiricists like Mill and Helmholtz, who claimed that the axioms of geometry formulated the laws of *space* established by observation; rationalists who alleged they could apprehend the "necessity" and "universality" of the axioms by considering the meanings involved; and Kantians who also found the axioms to be apodeictic but interpreted them as synthetic *a priori* propositions.

Poincaré was not a formalist in his approach to mathematics and he did not subscribe to the thesis that pure mathematics was simply a branch of logic. Nevertheless, he saw clearly that the axioms of a pure geometry were not propositions which could be affirmed as either true or false. For the terms they contain, like "point," "line," and so on, have no independent meaning or reference, and are "defined" and made precise only by the stipulations laid down for them by the axioms. A "point," for example, is anything which satisfies the conditions required by points by the axioms, so that nothing which fails to satisfy these conditions can be a point, and nothing can fail to be a point if it does satisfy them. Consequently, until a fixed reference to some *physical body or relations of bodies* is established for symbols such as "point," the axioms have no determinate subject matter and are not *propositions* capable of being decided as true or false.

It is important to note that a correlation must be established between the terms of geometry and *physical* configurations, and not between these terms and properties or elements of "space." We do not intuit "space" and it is not the object of observation; it is bodies and their traits which are observed. "Space" is regarded by Poincaré as a completely amorphous medium with no structure of its own, and we cannot therefore ask whether "space" is Euclidean or not; we can only ask whether a physical manifold is Euclidean or not, but only after we have *selected* certain *objects* which we have decided to *call* "points," "lines," and so on. As Poincaré says,

> We know rectilinear triangles the sum of whose angles is equal to two right angles, but equally we know curvilinear triangles the sum of whose angles is less than two right angles. The existence of the one sort is no more doubtful than that of the other. To give the name of straights to the sides of the first is to adopt Euclidean geometry; to give the name of straights to the sides of the latter is to adopt the non-Euclidean geometry. So that to ask what geometry it is proper to adopt is to ask, to what line is it proper to give the name straight? [120]

In effect, therefore, the adoption of Euclidean axioms is the adoption of a complicated "definition" for specifying the sort of objects we shall *call* "points," "lines," and so on; that is, those physical configurations are to be called "points," etc. which have the properties and relations specified by the formal system of geometry.

It may be thought, perhaps, that it might be possible to establish by a series of measurements whether a given manifold is Euclidean or not, without first having to employ a system of geometry for specifying what are points, lines, and so on. Poincaré argues that this alternative is out of the question. In the first place, measurements can be performed only *upon* physical objects and *with* physical objects; we can make no measurement or experiments upon "space." In the second place, since we must use instruments to measure, we cannot determine whether these instruments conform to Euclidean requirements by further measurements, without involving ourselves in an infinite regress. At some point in the process of instituting measurement, therefore, we are compelled to introduce a "convention," which stipulates which objects shall be *designated as* "straight lines," etc. But if we do not employ a system of formal geometry to specify this convention, what properties of objects can we take by means of which they can be reliably and repeatedly *identified* as points, lines, and so on? And in the third place, it is notorious that measurements do not uniformly yield precisely the same results; if we were to employ a different method for specifying the convention than that provided by a system of formal geometry, propositions dealing with the spatial relations of things will require constant revision, and sciences like mechanics which presuppose them would not be capable of the sort of development they have in fact received.[121]

One phase of Poincaré's conventionalistic thesis has now been outlined. But more must be said to give an adequate picture of his position. Poincaré recalled the dualities between the several pure metrical geometries which had been exhibited by Klein, and claimed as a consequence that there is no fact of experience expressible in terms of the distinctions of Euclidean geometry which cannot be stated in any of the non-Euclidean systems, and conversely.[122] Indeed, the only difference which could result is that things covered by one name on the first alternative would bear different names on the second, and would occupy different *logical* positions in the two systems. For Klein showed that the systems of abstract metrical geometries are intertranslatable. Consequently, a configuration designated as a "straight line" when one of these systems is employed as a conventional schema, will have different properties than the configuration designated by the same name when an alternative system is so employed. However, because of the dualities between the systems, there will correspond to what is called a straight line in the first case some configuration with the same properties in the second, though it will be designated by a different name.

Poincaré therefore concluded that the choice of geometry for physical investigations is not dictated by the facts of physics, but is the outcome of a convention. Nevertheless, while not *dictated* by the facts of physics, the selection of a geometry for "defining" the measurement of extensive features of objects is not *arbitrary:* the selection will be controlled by the requirement that the system of physics be the "simplest" and "most convenient" theory of natural processes which can be achieved. The evaluation of the appropriateness of a geometry cannot be made without considering geometry as part of the body of physical knowledge. That is why it is "meaningless" to ask whether Euclidean geometry is true. "As well ask whether the metric system is true and the old measures false; whether Cartesian coordinates are true and polar coordinates false. One geometry cannot be more true than another; it can only be *more convenient.*"[123]

Poincaré thus offered a new interpretation and a partial justification of traditional view-points. Pure or abstract geometry is *a priori,* and can be developed independently of the facts of sense experience; rationalists are correct in insisting on this. But the "axioms" of abstract geometry are not propositions—they are simply logical *frames* out of which propositions may be constructed. Hence the axioms are not "necessary," since alternative axioms may be formulated; when suitably interpreted to refer to a specified set of physical configurations, they yield propositions whose trust must be determined empirically. In this sense the empiricists are correct. On the other hand, if the "axioms" are taken as "implicit definitions" of the objects which are designated as "points," "lines," etc., the axioms cannot be proved or disproved by experiment, because every experiment involving measurement will presuppose the geometry in question. In this sense the axioms are *a priori,* and the Kantians are correct in their contention that we must employ organizing principles not capable of being *disproved* by experience. Nevertheless, each of these philosophies requires important qualifications. Each of them is mistaken in maintaining that *space* is Euclidean, because geometry as a branch of natural science has as its subject matter bodies and their relations, and not space. And each is mistaken in dogmatically assuming Euclidean geometry as the only set of principles capable of organizing physical measurement, because alternative metrical systems can be employed for defining the bodies to be designated as "points," "lines," etc., and for correlating the measurements upon stipulated characters of different physical configurations. The history of nineteenth-century geometry thus terminates in the view that while on the one hand geometry as a pure deductive discipline is *a priori,* because it is essentially a calculus engaged in the transformations of symbols, on the other hand geometry as an applied discipline is to be judged by the requirements of fact and convenience.

5

The segment of the history of geometry which has been considered in this essay provides ample material for reflection upon the role of mathematics in the enterprise of human knowledge, and upon the status of pure mathematics in general. The different techniques which have been developed in that history and the principles which came to be formulated in order to direct and coordinate research, are tools taken for granted today both by mathematicians and logicians. The distinction between a pure and an applied mathematics and logic has become essential for any adequate understanding of the procedures and conclusions of the natural sciences. Familiarity with the techniques of implicit definitions of terms and the method of their constructive explication is of equal importance for comprehending scientific method and contemporary discussions of it. And the concepts of structure, isomorphism, and invariance, which have been fashioned out of the materials to which the principle of duality is relevant, dominate research in mathematics, logic, and the sciences of nature. These techniques, principles, and concepts have been refined and extended by contemporary writers, and an essay much longer than the present one would be required to outline the improvements which have been made upon these intellectual tools and the conclusions which have been reached with their help. This passing reference to recent researches must suffice on this occasion. But a few general comments on the history that has been traced are both unavoidable and appropriate.

1) A striking and characteristic feature of the history of geometry during the nineteenth century is the increasing "abstractness" of its language and the progressive formalization of its procedure. This trait

has sometimes been regarded as a fault, by philosophers as well as mathematicians, and it has been said that the attempt to establish geometry as an autonomous discipline without reference to the observable traits and operations of the "natural world" has terminated in making of that ancient discipline something unintelligible or at best something trivial. However, the history of geometry which has been traced should help in making the tendency toward formalization at least understandable.

One of the first steps in the process of getting knowledge is the *symbolization* by means of counters of some sort (images, thoughts, words, diagrams, algebraic signs, etc.) of certain recurrent features of the environment. The symbols we employ, unlike most of the things symbolized, can be manipulated freely, because their construction and combination are in a large measure controllable by us. We learn to control the flux of things only on the basis of ascertained interconnections between present and absent existential traits. But such interconnections can be studied in detail, in safety, and at out leisure, only insofar as we can survey them dialectically—that is, *symbolically*. That is why, if knowledge is not to be identified with its object, knowledge is a matter of constructing, using, and coordinating symbols. We can represent symbolically existential traits not present, even if they are not accessible directly; and in order to achieve theoretical knowledge, we must learn to formulate or represent symbolically connections between characters of existence not directly accessible, and characters of existence directly present or capable of being brought into such presence. It follows that if the order of *discourse,* that is, of symbols and their interconnections, is to be adequate for expressing and ultimately for controlling new and unexpected natural traits, it must be developed on as comprehensive a scale as we know how. Discourse, language, our system of symbols, must be flexible enough and general enough to permit us to map out known and unknown routines and their terminations in nature, in anticipation of their actual occurrence. Consequently, a system of discourse which is constructed for representing exclusively a small segment of nature only, may become a handicap in the further exploration of nature. For symbols which refer exclusively to particular states of affairs cannot enter freely into various combinations with other symbols. The assigning of fixed, determinate references to symbols is often an impediment to successful inquiry rather than an advantage.

The history of geometry clearly illustrates and supports these dicta. As long as terms like "point," "line," "congruent," and so on, were associated with very specific configurations and relations of bodies, the symbolic operations of geometry were restricted by the conditions arising from the assumed references of the symbols. These conditions set the limits to the significant combinations of symbols in the calculus of geometry. Now while some limitation to the possible permutations of symbols is essential and desirable, it is not essential or desirable that the conditions relevant for certain *special* uses of a symbolic system should hinder the development of that system considered simply as a tool capable of any number of different uses. The symbolic operations of the traditional geometrical calculi could not all be performed in a uniform and general manner, and "exceptional" cases had to be recognized because of the associated references of its terms. When therefore the demands for symbolic operations which could be performed in a perfectly general manner were satisfied, and "geometrical" calculi were constructed which obeyed different though analogous rules of operation than those which obtained in traditional geometry, the terms which occurred in the new calculi were necessarily freed from their ancient meanings. And when this process of liberalizing the conditions for combining the symbols of a system was carried far enough, the only "meanings" which could be assigned to the terms were those which are involved in the operations and relations into which they could enter. The terminus of this process of liberalizing the conditions for the significant combination of symbols is the construction of purely formal or abstract calculi. And it should be noted that even if we think of the formal systems of mathematics with an eye to their *use* in the sciences, their field of usefulness and range of application is increased rather than diminished by constructing and developing such symbolic systems in abstraction from the possible applications of which they may be capable. One lesson of the history of mathematics seems to be that the efficiency and power of calculi is improved by loosening the associations between their terms and determinate particular states of existential subject matter.

2) Although the subject matter of traditional Euclidean geometry consists mainly of various relations between measures of lengths, areas, and volumes, the technical development of geometry as a demonstrative discipline has forced that subject matter into the back-

ground, and in some of its developments not only metrical notions but also spatial ones have been eliminated. What remains as the subject matter of pure geometry is its own symbolic operations. Few will deny, however, that the formal operations of geometry have been suggested by familiar relations exhibited in the physical world and by our own manipulation of physical objects. It often seems puzzling, therefore, why reference to such existential matters should on principle be excluded in the cultivation of abstract geometry, or why anyone should insist that the transformations of symbols in accordance to specified rules are the sole content of pure mathematics.

The history of geometry which has been sketched should, however, help remove the sense of discomfort which many experience with such conclusions. For deductive geometry, whatever else it may be, is a symbolic system, an instrument instituted to achieve definite objectives; like other instruments, it may be studied as an object in its own right, and alterations may be introduced into it which have no direct reference to the initial objective for which it was constructed. As a consequence, it may, and in fact does, acquire a power and scope which remove it from its original province. It may even happen, and in fact it does, that there is no existential continuity between the different fields to which it is applicable, so that it becomes impossible to characterize pure geometry in terms of an identifiable subject matter other than that of its own operations.

The formalistic conception of geometry is therefore not primarily an attempt to *dispense* with a subject matter for geometry, but to *identify* its subject matter in such a way that however "abstract" and "abstruse" the discipline becomes, the existence of that subject matter cannot reasonably be put in doubt. And whatever else may be said of the subject matter of demonstrative geometry, it does seem unquestionable that it engages in the combinations and transformations of "symbols" in accordance with prescribed rules of operation. No formalist need deny that the symbolic systems *may* acquire "meaning" by bringing them into connection with the terms of everyday discourse and with the behavior of things. He may even insist with Santayana that "the great glory of mathematics is to be useful while remaining free," and that if it were incapable of application it might be an innocent amusement but "a wasteful and foolish exercise for the mind." Nevertheless, he could safely claim that the interpretation of his symbolic systems can be prescinded in the actual work of mathematicians to the definite advantage of the discipline, and that the evaluation of mathematical contribution does not wait upon such interpretations being actually given.

3) The "conventionalism" which is the partial outcome of modern geometrical research has often been interpreted so as to make nonsense of the efforts of men to determine the objective order and connection of natural processes. In the name of a "conventionalistic" standpoint, formulations of existential interconnections which have been won with great effort and ingenuity and which serve as instruments of control of the flux of events, have been relegated to the domain of "fictions," merely human inventions, and "concealed definitions." It is understandable, therefore, why so many thinkers should find in any hint of a "conventionalistic" interpretation of theoretical science the promise of nothing other than absurdities and unintelligibilities. Nevertheless, in the light of the history which it terminates, a carefully formulated conventionalistic thesis turns out to be a reasonable doctrine. Without attempting such a formulation in this place, some of the clarifications which a conventionalistic analysis introduces into scientific procedure must be briefly mentioned.

Conventionalism calls attention to the fact that different conceptual or symbolic systems may be intertranslatable, because certain formal symmetries characterize them. If two such systems are formally identical and their elements can be made to correspond to one another in a suitable manner, it clearly is a matter of choice, not controlled by the traits of the subject matter to which the systems are both applicable, which one of them is to be employed for formulating connections between these traits. Two such systems cannot differ in their expressive force or internal logical structure. They can differ only in the notations they employ, and in the *logical* positions which corresponding elements occupy in each. It is therefore one of the services of conventionalism to have called our attention to the *medium* or system of symbols which formulates the orders of nature as ascertained in the natural sciences. As a consequence, our conceptions of what knowledge is are clarified, and the ground is prepared for isolating those factors in our theoretical formulations which refer to existential traits of a subject-matter, from those factors which minister to needs of convenience and economy.

More specifically, conventionalism calls attention to the methodological resolutions which control inquiry and which serve as organizing principles in the construction of theoretical systems. It emphasizes the selective activity that is required for achieving any knowledge, and so assigns a fundamental role to the physical activities of the knowing organism in the process of obtaining knowledge. Conventionalism therefore contains an implicit critique of the classical rationalist view that the sciences are engaged in disclosing immutable connections between substances each of which owns an inherent and intrinsic "essence" or "nature." It indicates the logical grounds for radical transformations of theoretical science such as has occurred in the transition from the Newtonian to relativity mechanics, and opens up possibilities for reconstructing systems of knowledge—possibilities that have been dogmatically rejected by those who view science as the direct grasp of necessary relations in nature. Finally, it introduces clarity into the way quantitative terms are employed in physical inquiry; for it insists that the procedures which are involved in using them, instead of being clarified in terms of allegedly ultimate categories of science or philosophy, themselves *clarify* and make intelligible everything else. Thus what "mass" or "temperature" means in physics is established then and only then when the methods by which masses and temperatures are measured are exhibited and described. Similarly, what "line" or "distance" means in physical geometry is also established, when the conditions which anything must satisfy in order to be *called* a line or a distance are specified by means of a determinate procedure. Conventionalism has its mite to offer in the continual struggle for clarity of understanding and integrity of belief.

10/ Determinism and Development

The word *development* is notoriously one with protean meanings. It is sometimes used to connote a process, sometimes the product of a process. It is frequently employed as a purely descriptive term to characterize several types of change; but it also functions in many contexts as a eulogistic label. An analogous though perhaps less disturbing plurality of connotations is associated also with the word *determinism*. Prefatory to discussing the ostensible theme of this paper, something must be said, if only in a loose way, to identify the senses in which these words are to be understood in the sequel.

1

Even if, as in the present paper, the term *development* is used to signify a temporal process rather than its product, a number of its more specialized meanings, or components in its meaning, require to be distinguished. In many of its current uses the word carries the suggestion that developmental processes make progressively manifest something latent or hidden, a suggestion that is reinforced if we recall the original meaning of the word as connoting an unfolding or unwrapping. We still speak of the development of heat in a wire, of the development of a photographic plate, of the development of a fertilized egg, or of the development of a human personality, understanding in each case a sequence of continuous changes eventuating in some outcome, however vaguely specified, which is somehow potentially present in the earlier stages of the process.

It is not difficult to appreciate why so many writers both past and present, who have perhaps been influenced by the etymology of the word, have been unable to conceive of genuine alternatives to preformationist theories of development, and why they have therefore rejected epigenesis as "incomprehensible." But in addition to this backward reference, the designation of a process as developmental also has a prospective one, as the above examples of usage indicate. No change per se is commonly counted as a developmental one, though it may be so labeled if it is referred to an explicitly or tacitly assumed consequence of the change. The word thus possesses a strong teleological flavor. The teleology intended does not, of course, assume the operation of purposes or final causes; the imputation signifies only that a sequence of change is designated as developmental only if it contributes to the generation of some more or less specifically characterized system of things or property of things.

But a still further component in the notion of development needs explicit mention. There are contexts in which the word is applied to systems undergoing cyclic or repetitive alterations—as when the periodic return of the seasons, or the rhythmic motion of the heart, is counted as a case of development. On the other hand, many students often decline to apply the term to merely repetitive changes. For example, the descent of the weight in a pendulum clock is frequently not regarded as a genuine developmental process, though the slow wearing away of the cogs which eventually results in the irremediable stoppage of the mechanism may be so classified. In this usage, changes must be cumulative and irreversible if they are to be labeled as developmental. Moreover, in a somewhat narrower sense, the term is reserved for changes which are not merely irreversible, or which yield only a greater numerical complexity; those changes must in addition eventuate in modes of organization not previously manifested in the history of the developing system, such that the system acquires an increased capacity for self-regulation, a larger measure of relative independence from environmental fluctuations. This is, at any rate, what embryologists and evolutionists have in mind when they take progressive differentiation and self-maintaining organization as the essential marks of development. (For reasons of space, the use of the word to denote "retrogressive" as well as "progressive" changes will be ignored.) In discussions of moral growth the word is employed in what is perhaps only an analogous sense. But here, too, an individual is commonly said to be developing only if he is progressively exhibiting greater sensitivity and coordinated response to various cultural stimuli, and if his responses and attitudes fall into a stable pattern that is adapted to the vicissitudes of external fortune.

The connotation of *development* thus involves two essential components: the notion of a system possessing a definite structure and a definite set of pre-existing capacities; and the notion of a sequential set of changes in the system, yielding relatively permanent but novel increments not only in its structure but in it modes of operation as well.

Although this account is patently imprecise, it must suffice for my purposes; and I must be even briefer in explicating the meaning of *determinism.* In the loosest relevant sense of this word, it is a label for the claim that all things, events, processes, and traits come into existence, endure, or pass out of it, only under fixed and definite conditions. This claim can be stated more precisely as follows: Let S be any system of things occuring in an environment E; and suppose that at an arbitrary time S is in a certain "state" A (i.e., the components of S, their organization, and the specific traits of S at that time are described by "A"). One part of the deterministic assumption then maintains that if at a later time S should be in state B, the change from A to B is a consequence of some alteration either in S, or in E, or in both. Moreover, let S_1 and S_2 be two systems occurring in environments E_1 and E_2 respectively; and suppose that S_1 changes from state A_1 to state B_1, while S_2 alters from state A_2 to state B_2. Then determinism also maintains that if B_1 and B_2 differ in some respects, this difference is the consequence of some difference either in the intrinsic composition of the two systems, or in the environments of the systems, or in the initial states of the systems, or in the changes initiating the transitions from the initial states, or in all of these together. If determinism is assumed, alterations in a system which do not appear to occur as the consequence of the presence or operation of antecedent factors or conditions, must be regarded as belongirg to a more inclusive system which is deterministic. On the other hand, since we are admittedly ignorant of the conditions upon which many occurrences are contingent, determinism in effect functions as a maxim that formulates the formal objective of empirical inquiry. Moreover, since in its general formulation the deterministic assumption does not specify upon which definite conditions a given type of occurrence is contingent, the content of this assumption is meager and cannot be unambiguously controverted by factual inquiry. Despite these difficulties, I wish nevertheless to discuss the import for determinism of two areas of inquiry in which developmental notions play important roles, one of the areas being biology and the other human personality.

2

On the surface at any rate, there seems to be no incompatibility between the findings of developmental studies and determinism broadly construed. For on the one hand, determinism merely expresses in general form that component in the meaning of *development* which connects the outcome of developmental process with antecedent structure and latent capacity. On the other hand, determinism does not deny that many systems manifest in their development novel traits and new levels of organization. For the deterministic formula is silent on the question whether traits exhibited at later stages of development are novel or not, nor does it prescribe the kind of materials and changes which must concur as the condition for the appearance of diversified structure and function.

Nevertheless, a closer look does reveal sources for the equivocal attitude of many students of biology toward determinism. For according to determinism, if in the parallel development of two organisms existing in identical environments, one exhibits traits which the other does not, the difference must be accounted for in terms of differences in the constitution of the two organisms, or in their initial states. As Driesch put it, determinism assumes that "no states and no events in nature are without a sufficient reason for their being such as they are at such a place and time," so that determinism implies the "univocal determination of being and becoming." To cite but one familiar example, however, it has apparently been established that despite an initial homogeneity, or equality, in the intrinsic composition of certain embryonic parts, those parts are eventually transformed into inhomogeneous, or unequal, parts of the adult organism. Such findings are often felt to raise difficulties for the deterministic assumption.

I am not qualified to discuss the factual issues raised by these embryological experiments. I do want to consider, however, the three main lines of the effort which has been made to save determinism in the face of such apparently fatal experimental findings. One of them is *vitalism,* according to which differences in development from allegedly identical antecedents must be explained in terms of the operation of non-spatio-temporal determinants of entelechies. But vitalism has not proved to be a fruitful notion, and it no longer seems to present a live issue in the philosophy of biology. A second defense of determinism is based on the hypothesis that the apparent equality of organic parts which develop differently is really only apparent, and that a fuller study will eventually reveal subtle but still obscure physicochemical differences between them. This is the standpoint of what is commonly known as *mechanism* in biology, a standpoint that has been rejected not only by vitalism but also more recently by so-called *organismic* biologists. According to the organismic approach, which is the third way mentioned of saving determinism, the mechanistic conception rests on a "machine theory" of living organisms, and subscribes to an *additive* notion of organic structure. Mechanism is therefore said to be in principle incapable of explaining vital phenomena in their totality, since the additive point of view cannot do justice to the hierarchically organized structure of living organisms, and cannot account for the distinctive traits which are manifested on each level of the hierarchy, or for the integrated, self-maintaining character of the whole organism. In substance, therefore, the organismic standpoint is a variant of the doctrine of emergent evolution, for like the latter it maintains that traits exhibited by a hierarchically organized system cannot be reduced to, or explained by, the properties of parts of the system whose mode of organization occurs on a lower rung of the hierarchy.

I cannot here discuss in detail the mechanist and organismic standpoints; I can offer only a few brief remarks toward evaluating their respectives claims, and I hope to show that both positions overlook important points in the logic of explanation. Let me begin by noting that systems can undoubtedly be distinguished from one another according to the degree in which they possess a self-maintaining or self-regulative character. This distinction can be made in an objective way, without introducing assumptions about purposive agents or self-realizing ends-in-view. Indeed, the distinction is entirely neutral to the difference between the living and the inanimate. Even systems whose detailed modes of operation can be completely understood in terms of the principles of classical mechanics may be self-regulative with respect to certain of their manifest traits. It is, therefore, clearly a mistake to claim that mechanistic explanations of living organisms are in principle impossible, merely on the ground that such systems are adaptive and self-maintaining.

In the second place, the question whether a given trait of a system, which it manifests as the outcome of a developmental process, is "novel" or "emergent," requires to be so formulated that the attribution of novelty is relative to some specific explanatory theory. Let *S* be some given system which is being studied with respect to a certain set of its "global" properties $P_1 \ldots P_n$; and suppose a number of laws or regularities $L_1 \ldots L_k$ have been established between some if not all of these properties. Suppose, moreover, that a comprehensive theory *T* has been devised which analyzes *S* into a set of parts $X_1 \ldots X_y$, and which is formulated entirely in terms of traits $Q_1 \ldots Q_j$, characterizing those parts. Assume, finally, that the theory stipulates certain relations between the *P*'s and *Q*'s so that when appropriate boundary and initial conditions are supplied, the occurrence of a number of the *P*'s can be predicted and a number of the laws *L* can be deduced. Two main cases can now be distinguished: all the laws can be deduced and the occurrence of all the *P*' can be predicted; or this can be done only for some of the *L*'s and *P*'s. In the first case, the system *S* with respect to the properties *P* and laws *L* is deterministic relative to the theory *T*; in the second case, *S* is only partly deterministic, and those of its properties and regularities which cannot be predicted on the basis of *T*, are emergent properties and regularities *relative* to *T*. Nevertheless, a different theory *T*' might be found, which is formulated in terms of an alternative set of parts of *S* and their traits, such that relative to *T*' the system *S* is deterministic in respect to its properties *P* and laws *L*. In short, whether a system is deterministic in respect to a given set of its features is a question that requires to be put relative to some specific set of theoretical assumptions.

In the third place, the question whether a system S is an additive one or not must be made relative in a similar manner. Continuing the above notation, suppose that a theory *T* can explain the occurrence of the properties *Q* of the parts *X* of *S*, when these parts occur in a certain set of relations *R*, but that it cannot do this when those parts stand to each other in some other set of relations, and in particular when they stand to each other in just those relations that constitute the organization of *S* itself. In this event, *S* is not deterministic with respect to the theory T, and it can be said to possess various "holistic" traits which are not reducible to the traits of its parts. On the other hand, some other theory T' may be invented, which is formulated on the basis of an alternative analysis of S into parts X' and their properties Q', such that T' does make it possible to predict the occurrence of the global properties P of S when the parts X' stand to each other in just those relations which constitute the structure of S itself. Accordingly, S can then be said to be an additive system relative to T', though not relative to T, so that whether or not the system is an additive or organic one depends on which theory is being assumed.

The possibilities just outlined are not introduced simply as a formal exercise. They represent schematically what I believe are actual situations in the history of science. For example, relative to Dalton's atomic theory, many of the properties of chemical compounds are emergents, and with respect to these properties chemical compounds are organic or nonadditive systems relative to that theory. But relative to current quantum theory, a number of those properties which are emergents for Dalton's theory no longer are emergents; and with respect to these properties chemical compounds are additive systems relative to quantum mechnics, though not additive relative to the older theory.

Let me now state briefly what I think is the outcome of this discussion. The main conclusion is that determinism is compatible with the known facts of experimental biology, not only if determinism is stated in its most general form, but also if it is made more specific by construing it to postulate physicochemical determinants for all occurrences. It is, of course, quite evident that if a mechanistic theory in biology does not include in its vocabulary, either as a basic or as in some sense a "defined" term, an expression which describes a distinctive outcome of a developmental process, that theory cannot possibly predict or explain the manifestation of such a feature; and if that feature is a novel one in the history of the developing organism, the theory cannot account for this novelty. All this is just a logical truism. But none of this can be taken to imply that an alternative theory could not possibly be devised, which would be mechanistic by common consent, such that it would contain the previously missing term in its vocabulary and so be in the logical position to account systematically for the novel property. Transformations of this sort which mechanistic theories have undergone in the history of science provide sufficient evidence to the contrary. On the other hand, there is no a priori proof that a mechanistic theory (or for that matter, a theory of a different type) can always be found which will explain the occurrence of what is assumed to be novel; nor is there plausible ground for the supposition that some day men will construct a complete and final theory which will account for every trait that may ever be manifested. The chief point upon which I am insisting is that the notion of novelty, as it enters into the meaning of *development*, needs to be explicated relative to a specific theory, and that there are no substantial grounds for the thesis that deterministic theories of the mechanistic type are in principle incapable of dealing adequately with developmental processes. Nevertheless, the question whether a mechanistic theory of biological development can in fact be constructed, and the question whether the cause of knowledge is advanced by current proposals of such theories, are issues whose answers depend partly on the character of biological subject matter, partly on the ingenuity of men, and partly on the strategy of research at a given stage of inquiry.

3

The influence of the belief in a universal determinism upon conceptions of human personality has been of acute concern to students of ethics and morals, and I would like in conclusion to touch briefly upon one of the issues that have been raised. There is today an extensive literature whose main burden is that as a consequence of the current extensions of causal analyses to human behavior, and especially as the result of seeking physiological and social determinants for individual actions, the possibility of autonomous personal development is being challenged, and the sense of personal responsibility is being gradually undermined. As one recent writer has put it, methods are being employed in studying men which are at best adequate for understanding the nature of machines and rats; and since in consequence men are being viewed simply as the products of the conditions to which they have been exposed, it has become fashionable to hold no individual responsible for anything that he does because the locus of responsibility has been shifted to the conditions which allegedly have made men what they are. And as another commentator has claimed, on the determinist view the notion of human responsibility as commonly understood is inapplicable to actual individuals. Eloquent pleas have therefore been made to cease viewing men as nothing but machines subject to deterministic laws, and to recognize that men are capable both of making choices which are not completely determined and of undergoing a free autonomous development.

It must be admitted that it is easy to find in current social and psychological discussions much that supports the fears which such complaints illustrate; and it is not my intention to support the amoralism professed by many contemporary students of the human scene. Nonetheless, it is far from clear that commitment to any form of determinism is incompatible with the facts of moral life or with imputations of individual responsibility. If it is the case that men engage in deliberation and make choices only under determinate conditions—whether these conditions be physical, physiological, psychological, or social—the discovery that this is so, and the identification of those conditions, do not, on pain of a radical incoherence in our thought, eliminate from existence those very features of human behavior which the discovered conditions for their occurrence ostensibly maintain. Should we succeed, for example, in ascertaining some day the particular collocations of physical particles and their modes of interaction, upon which the occurrence of self-consciousness and acts of deliberation are contingent, we should not thereby have explained away those traits of human life that initiated the inquiry into their physical conditions. The assumption or the discovery that our acts and choices are determined in some fashion does not mean that we are being coerced when we are engaged in deliberation and decision, nor does it mean that acts of deliberation and choice are irrelevant to what we may overtly do. But the mere absence of feelings of coercion does not itself warrant the conclusion that there are no determinants, whether physical or otherwise, for what we are and what we do. In short, the assumption that responsible choice and action are manifested as products of developmental processes which have conditions for their existence, does not convert the moral life of man into a sham and illusion.

Nor does the deterministic assumption in any of its general or specialized forms necessarily place limits upon legitimate human aspirations or undermine the grounds for imputations of personal responsibility. On the contrary, the discovery of the conditions under which various distinctive human traits occur may itself become the means for releasing or redirecting human energies. This does not deny that a specific causal hypothesis—for example, one which locates the determinants of human action exclusively within the epidermis—may be inadequate. But it is well to bear in mind that even in purely physical inquiry a theory can account for the actual operations of a system only if it is supplied with the requisite initial and boundary conditions, which may be variable for different systems; and two systems whose operations are intelligble in terms of the same general principles may nevertheless behave differently if their special structures and the special circumstances of their operation are different. In any event, as the practical problems of education and the administration of justice clearly indicate, an intelligent imputation of responsibility involves the question of where the factors are located that are simultaneously controll*ing* in human action, and also controll*able* by human intervention. We may be right or we may be mistaken in supposing that these factors are of one sort rather than another; but the question cannot be settled without detailed inquiry or by invoking the alleged autonomy of human choice. There undoubtedly are commentators on the human scene who are distressed by the suggestion that we must engage in empirical inquiry in order to establish the limits of individual responsibility, just as there are temperaments that are revolted by the idea of botanizing on one's parent's grave. But if our objective is really to know and find things out, the patent facts of moral experience present an invitation for discovering the conditions, whatever these may be, under which growth of human personality is possible, and under which the scope of moral choice may be enlarged.

11/Some Notes on Determinism

Whether the occurrence of every discriminable event is determined, whether for every event there is a unique set of conditions without whose presence the event would not take place, and whether if conditions of a specified kind are given an event of a certain type will invariably happen, are variant forms of a question that cannot be settled by a priori arguments. Nor do I think the question can be answered definitively and finally, even on the basis of factual evidence; for, as I shall presently suggest, the question is best construed as dealing with a rule of procedure for the conduct of cognitive inquiry, rather than with a thesis concerning the constitution of the world. I am therefore not convinced by Professor Blanshard's acute argument attempting to show that an answer to the question other than an affirmative one is indefensible, if not unintelligible.[1] Moreover, his assertion that even in deductive thinking and artistic invention each step is necessitated by the logical and aesthetic relations that exercise a power over the mind seems to me untenable—if it is admitted as relevant to his major contention. For his assumption that logical and aesthetic relations (as distinct from *apprehensions* of such relations) can engage in causal action attributes causal efficacy to things that, in no recognizable sense of the phrase "causal action," can exercise such agency.

Nevertheless, the belief in determinism is not unfounded; and it would be just silly to maintain that in no area of experience can we rightly affirm that anything is caused or determined by anything else. It seems to me the special merit of Professor Black's paper[2] that it indicated at least one identifiable class of contexts in which the words "caused," "determined," and their derivates have an unquestionable and important use. In these contexts involving human action it is simply nonsense to deny that events have causes or effects, in senses of these words appropriate to these contexts. Black has also made plain that the conditions under which "caused" and "determined" have significant uses, in those situations where men initiate actions and are responsible for the occurrence of events, require the presence of identifiable contingenices and the absence of just such "necessities" as those for which Blanshard argues.

But it does not follow from Black's analysis that the only sense that can be attached to "cause" and "determined" is the sense they manifestly do have in the indicated contexts—any more than it follows that, because the word "number" is undoubtedly used in situations involving the counting of objects, the meaning of the word in statements about such irrational magnitudes as the area of a circle with a unit radius must also involve reference to counting. There are, to be sure, historical continuities and important anlogies between the use of these words in contexts of human action and their use in discussions about, say, the "indeterminism" of electrons. But it is patently a mistake to construe the meanings of those words in this latter context in terms of the "paradigm" for their use in situations where men are correctly identified as causal agents. Although Black does not explicitly guard himself against the suspicion that he does take his paradigm as basic for all uses of "caused," it is unlikely that he would commit himself to such a position. On the other hand, it does seem to me that Professor Bridgman (and perhaps even Professor Landé) commits a somewhat similar mistake when he suggests that the "indeterminacy" of quantum theory can be explicated in terms of familiar facts "in the sphere of ordinary life."[3]

In the voluminous literature on the "indeterminism" of microphysics, one point stands out clearly: whatever the issue may be, it is generated by the theoretical interpretations that are placed on the acknowledged data rather than by any disagreement as to what those data are. Thus no one disputes that when a beam of light passes through appropriately arranged slits and strikes a zinc sulphide screen, scintillations occur that fall into a definite pattern; or that quantum theory accounts admirably for the occurrence of this pattern; or, finally, that there is no known law of physics that accounts for the occurrence of each individual scintillation. Problems arise, however, when the structure of quantum theory is analyzed with a view to showing why it is that this theory cannot account for individual scintillations. But the problems are generated because answers to the question are proposed in terms of familiar facts "in the sphere of ordinary life" rather than in terms of the structure of the theory itself.

It is a commonplace that quantum theory employs a distinctive way of "describing" the state of a physical system with which the theory can deal. This state description (the psi function) is such that, given its value for some initial time, and assuming an appropriate set of boundary conditions for the application of the theory, the theory makes it possible to calculate the value of the function for any other time. In this respect quantum theory is as "deterministic" as are the dynamical theories of classical physics. It differs from these in that, while the state description of the latter can be construed as representing magnitudes associated with certain individual elements that constitute the physical systems in question, its state function can be construed as representing only a statistical property of the individual elements making up the physical system. In short, the state description of quantum theory is a statistical parameter. So far nothing could be more straightforward or less puzzling. The puzzle begins when reasons are offered why the state function of quantum mechanics is a statistical parameter.

The reason Professor Landé appears to give is that any given value assigned to the psi function constitutes the initial conditions for the application of the theory to a concrete situation, and that since initial conditions constitute a brute and underived fact they represent an inherently chance or random feature of the world.[4] I doubt very much whether I have understood Professor Landé's presentation of his views, and my comments may be entirely irrevelevant to his real intent. But as I understand him, he has not made clear what he set out to clarify. For every theory—not only quantum mechanics—requires initial conditions that at some point or other in an investigation into concrete subject matter must be accepted as underived and therefore as representative of a "random" feature of the world, as are the initial conditions for quantum theory. This attempt to assimilate the "indeterminacy" of microphysics to facts "in the sphere of ordinary life" is not a successful one.

Professor Bridgman seeks to explain the statistical aspect of the psi function by invoking the general principle that whenever measurements are made the instruments employed interact with the things measured and thereby introduce changes into the latter. His contention is that, although such alterations are practically negligible when we measure things that are sufficiently large, the changes cannot be ignored when the minute "particles" of microphysics are measured with the relatively large instruments at our disposal, so that the psi function inevitably represents only statistically significant magnitudes associated with the elementary particles of quantum physics. Now, the general principle Professor Bridgman invokes is undoubtedly sound. The difficulty in his explanation, however, is that though the principle is sound it does not, in other areas of inquiry, prevent us from calculating the effects of measuring instruments on the things measured and so making corresponding allowances in assigning magnitudes to the objects under investigation. Why should the situation be inherently different in quantum physics? I find it difficult to escape the impression that Professor Bridgman has put the cart before the horse. For it seems to me that the alleged effect of measurement on microphysical "particles" must be assumed as at best a consequence that follows from the acceptance of quantum theory, rather than that the theory is based on independently ascertained facts concerning the alterations made by instruments of measurement on microphysical "particles." At any rate, I do not think Professor Bridgman has convincingly shown that the "indeterminism" inherent in the structure of quantum theory is but another illustration of a familiar feature "in the sphere of ordinary life."

Although quantum theory is not deterministic in the precise sense in which the dynamical theories of classical physics are deterministic, and although quantum theory (or, for that matter, any other available theory of contemporary physics) does not account in detail for such occurrences as the individual scintillations mentioned previously, it of course does not follow that there really are no precise conditions for the occurrence of those events that quantum mechanics does not explain, or that a theory that can account for these things is impossible. The assumption that there always are such precise conditions for every event, even if we continue to remain permanently ignorant of them, is the assumption of a universal determinism. As I have already indicated, determinism so understood is capable neither of decisive proof nor disproof. I think, nevertheless, that determinism can be regarded as a fruitful maxim or regulative principle for inquiry. It does not express a necessity of thought, for it can be abandoned. But if it is abandoned, then inquiry in certain directions is, at least temporarily, brought to a halt. In an important sense, therefore, the deterministic maxim is explicative of what is generally understood to be a goal of the scientific enterprise.

I want to conclude with a brief comment on the contention of Professors Edwards and Hospers that, if determinism supplies a true account of the nature of things, it does not make sense to hold anyone morally responsible for his actions or to offer moral praise and blame.[5] Under what conditions do we hold a person morally responsible for an action? Consider an example. I engage the services of a student as baby sitter on the assumption that she is capable of doing certain things. Her ability to do them depends on a number of conditions, including the state of her body, her education, and her previous experience with children of a certain age. If she does indeed satisfy these conditions and also agrees to perform certain tasks that are compatible with her abilities, she is morally responsible for performing them. The fact that she did not create her own body, or that she did not choose the education she received, are not relevant considerations for judging whether she is morally responsible for some event that may take place during my absence from home. On the other hand, if during my absence burglars enter and tie up the student, or if she becomes unconscious for causes not within her control, she is not morally responsible for what may befall my children. The point is obvious. Moral responsibility is correctly ascribed to individuals who possess certain capacities; and it is correct to make the ascription for the sufficient reason that this is just the way the phrase "morally responsible" is used. The fact that possessing these capacities is contingent on a variety of conditions, most of which are perhaps beyond the control of an individual, is irrelevant to the analysis of what we do mean by the phrase as well as to the grounds on which the ascription is rightly made. To maintain the contrary is in effect to maintain that no property can be correctly predicated of an object if the property is causally dependent on anything either in the composition or in the environment of the object. Such a view makes all predication impossible. But in any event Professors Edwards and Hospers can sustain their thesis only by radically altering the customary conception of what it means for anyone to be morally responsible.

12/Teleology Revisited

A. Goal-Directed Processes in Biology

"Naturalism" is the label for a number of distinct though related philosophical doctrines; but this is not the occasion for listing their varieties or identifying what is common to them. It does seem appropriate, however, to begin these lectures with a reminder of the sense in which John Dewey characterized his logical theory (or theory of inquiry) as "naturalistic." He so described his theory because in it "there is no break of continuity between operations of inquiry and biological operations" as well as physical ones, and because he believed he had succeeded in showing that the "rational operations [of inquiry] . . . grow out of organic activities, without being identical with that from which they emerge."[1] This summary account of Dewey's conception of naturalism is not informative about the nature of the "continuity" he sought not to breach. However, it does make evident that for him biological processes and biological theory constitute a substantial part of the matrix in which his theory of logic has its roots. Dewey also maintained that a touchstone of whether Darwinian ideas (as he understood them) concerning logical method have been properly understood and assimilated, is the way the much-debated problem of "design *versus* chance" is treated.[2] The conspicuous place this problem and its treatment occupied in his assessment of the significance of Darwinian theory for philosophy testifies to the importance Dewey attached to the roles teleological notions play in biological inquiry.

Dewey's interest in the issues raised by the use of teleological conceptions is *one* reason for my choice of teleology as the theme of these Dewey Lectures. However, the debt owed to piety is not the *only* reason. Although the ideas associated with the term "teleology" have had a long and checkered history, and teleological language continues to be viewed with suspicion by many scientists and philosophers, its use in biology as well as in the psychological and social sciences is widespread. Indeed, many biologists believe that such language is indispensable for describing and explaining a large variety of important biological phenomena. But despite the voluminous literature devoted to analyzing teleological concepts that has appeared in recent years at an accelerated rate, there is much disagreement both over the meaning of teleological statements and concerning the kinds of events and processes about which it is appropriate to assert them. These disagreements are not easily resolved for a variety of reasons. But whether or not they can be resolved at all, it can be instructive to reexamine some of them in the light of fresh evidence and new perspectives on the issues at stake; and I want in these lectures to do so, though restricting myself mainly to questions that are germane to biology. The recent literature sometimes breaks fresh ground; but much of it consists of modified (and sometimes much improved) versions of well-known views, presentations of difficulties in older analyses, or challenges to basic assumptions underlying customary approaches. My discussion will, in consequence, inevitably revisit much familiar territory, though with some different objectives than on previous journeys.

1

A commonly recognized but loosely delimited trait of biological organisms, a trait that is often said to distinguish living from inanimate things, is the apparently purposive character of living organisms. Teleological language reflects this distinction. Although there is no exhaustive set of criteria for distinguishing teleological locutions from those that are not, the occurrence of certain expressions in statements is usually a fairly reliable indication that the statements are teleological. For example, the statements "The purpose of the liver is to secrete bile," "The function of the kidneys is to eliminate waste products from the blood," and "Peacocks spread their tail feathers in order to attract peahens," are teleological; and the occurrence in them of the expressions "the purpose of," "the function of," and "in order to" is a sure indication that this is so. However, these are not the only expressions whose occurrence in a statement marks the latter as teleological. There are also statements that are generally acknowledged to be teleological, even though they contain no typically teleological phrase—for example the statement "The hare sought refuge in its burrow from the packs of pursuing hounds." But despite the absence of a formal criterion, there is little disagreement in actual fact as to whether a given statement is teleological or not.

Teleological statements are not all of the same kind, and they can be classified in a number of ways. One distinction that will be useful in what follows is between "goal ascriptions" and "function ascriptions." The former state some outcome or goal toward which certain activities of an organism or of its parts are directed. For example, each of the following is a goal-ascription: "The goal of the pecking of woodpeckers is to find larvae of insects," and "The goal of the activities in various animals of the sympathico-adrenal apparatus as well as of certain cells in the pancreas is to keep the concentration of blood sugar within relatively narrow limits." On the other hand, function ascriptions state what are some of the effects of a given item or of its activities in an organism. For example, the following statement is a function ascription: "The function of the valves in the heart of a vertebrate is to give direction to the circulation of the blood."

Some biologists use the words "goal" and "function" interchangeably, ignoring distinctions that linguistic usage recognizes—they do so possibly because the distinctions are not relevant to the tasks in which they are engaged.[3] But however this may be, *seeing* is customarily said to be a function of eyes, rather than their goal; and *escape from a predator* is said to be the goal of a hare's flight, rather than its function. Moreover, although the end-products of certain goal-directed processes have a function (for example, one function of the homeostasis of sugar concentration in the blood is prevention of convulsion), this is not always the case (for example, survival may be the goal of a hare's flight from a hound, but survival itself does not appear to have any function). Also, the lachrymal glands have a function, namely to lubricate the exposed surface of the eyeballs, but do not have a goal. But the main reason for retaining the distinction is twofold: the analysis of what constitutes goal-directed behavior is different from the analysis of what counts as a function; and secondly, the structure of functional explanations in biology differs from the structure of explanations of goal-directed behavior. The present lecture will be devoted to the discussion of goal-directed processes in biology, and the second one to functional explanations.

2

There are a number of alternative (but not necessarily incompatible) explications of the notions of goal and goal-directed processes, and I will examine three of them. Undoubtedly the most familiar account of these notions takes the primary or "core" meaning of these terms to be determined by their use in connection with purposive human behavior (and perhaps also in connection with the supposedly purposive behavior of the higher animals). Accordingly, the "goal" *G* of an action or process is said to be some state of affairs *intended* by a human agent; the *intention* itself is an "internal mental state" which, coupled with the internal state of "wanting" *G* together with "believing" that an action *A* would contribute to the realization of *G*, is allegedly a causal determinant of the ensuing action *A*. "Goal-directed" behavior is then the action *A* undertaken by the agent for the sake of achieving the goal. For convenience of reference I will call this account of goals and goal-directed processes the "intentional" view.

Several features of this account should be noted. In the first place, though the *occurrence* of the action *A* can be explained teleologically, the explanation is ostensibly a species of *causal* explanation. For by hypothesis, the action is initiated because the agent *desires* a certain goal and also *believes* that the action will contribute to its production.[4] The causal explanation cannot be rightly charged with the difficulty often raised against teleological explanations, that the causal explanation assumes that a future state of affairs can be causally efficacious in bringing about its own realization. For according to that explanation, it is not the *goal* that brings about the action. It is rather the agent's *wanting* the goal, together with his *belief* that the action would contribute to the realization of the goal, that does so.

In the second place, this account is fully compatible with the usual characterizations of goal-directed behavior. For example, human beings are said to be goal-directed even when they fail to achieve their goal; and this is congruent with the intentional view, according to which an action is goal-directed if it is undertaken for the sake of some *intended* goal, whether or not the goal is reached. Again, goal-directed behavior is said to be "plastic," in the sense that the goal can be reached by alternative routes and from different initial positions, the route or action that is actually adopted depending on what local circumstances prevail. But such plasticity is obviously compatible with the intentional view of goal-directed behavior, for the alternative actions that are recognized as possible roads to the goal, as well as the action that is actually adopted, will depend on the beliefs which the agent has in the situation in which he finds himself concerning the available means for reaching the desired end.

On the other hand, if the intentional view of goals is taken literally, an organism can be described as goal-directed with respect to a determinate goal *only if* it is legitimate to ascribe intentions, desires, and beliefs to the organism. In consequence, goals and goal-directed behavior can be correctly predicated only of human beings, and possibly of some higher animals. It is therefore entirely inappropriate to use such language in connection with organisms such as protozoa and plants which are incapable of having intentions and beliefs; in connection with subsystems of organisms, such as the complex of glands and other parts of the human body, that are involved in the homeostasis of the blood temperature; or in connection with inanimate systems, such as a steam engine provided with a governor and other servomechanisms that are sometimes described as goal-directed.

To permit the ascription of goals in some of these cases as well, some proponents of the intentional view have broadened the notions of goal and goal-directed behavior. According to Dr. Andrew Woodfield, for example, this more inclusive notion of having a goal is an *extension* of the "core concept" of having a goal—that is, the concept of an intentional object of desire—to systems possessing "internal states" *analogous* to the internal state of wanting a goal.[5] The goal-directedness of servomechanisms, he maintains, "consists in the fact that they behave as if they had desires and beliefs in virtue of the fact that they are feedback systems. . . . [T]he 'desired' end-state is encoded in their internal structure."[6] If an internal state of a machine is to be described as having the goal *G*, the internal state must not only cause the machine to behave in a manner that leads to *G*, but it must also "represent" *G*. However, Woodfield continues, "[t]here are no clear rules for deciding when an internal state is sufficiently similar to a desire to count as the state of 'having a goal.' "[7] Nor does he attempt to state necessary and sufficient conditions for an internal state to count as a representation of the goal. He does not think, for example, that an electronic feedback-subsystem, which maintains a steady output of electric current in the main system, is an internal state of the latter sufficiently similar to a desire to count as the state "having a goal." Since there appears to be no significant material difference between such a system and either physiological homeostats or artificial thermostats, he excludes both of those from the class of goal-directed systems.[8]

It is a plausible claim that the primary meaning of the phrase "being goal-directed" is stated by the intentional view; and it is not unreasonable to suppose that the prevailing more inclusive sense of the phrase is a "metaphorical extension" of its initial meaning. In any case, since biological inquiry deals for the most part with matters in which desires and beliefs do not occur, it is obvious that were the phrase used strictly in the sense specified by the intentional view, it would have no application in most parts of biology. However, if the notion of having a goal that is not intended by anyone is held to be an impossibility, the sole consequence that this definitional resolution would most likely have is that a new word would be coined for the steady states achieved by physiological homeostats. Moreover, it seems to me dubious that there must be close analogies, as Woodfield maintains, between the items involved in the primary sense of the term "goal" and those involved in its more inclusive current usage. Nor does his requirement that there be such analogies help to explicate the sense of goal-ascriptions in biology. Consider his claim that the inner state representing a goal in an ostensibly goal-directed biological process (for example, the process in which a tadpole develops into an adult frog) must "resemble" the inner state representing a goal that is pursued in some purposive human behavior (for example, the process in which a person seeks to recover a coin that fell into a crevice). In the latter example, the inner state according to Woodfield is a complex mental state involving an intention, a desire, and a belief. But in what way does the inner state representing the goal in the former example—an inner state which is perhaps a complex subsystem of genetic materials in the tadpole—resemble, or is analogous to, a desire or a belief? It is not clear that the question really makes sense, and in any event it is difficult to know how to begin making the relevant comparison. Moreover, even if this hurdle is jumped and it is assumed that the goal-directed development of the tadpole might be analogous to the goal-directed efforts of human beings, the analogy could surely not be established by comparing the *inner* states themselves. What might be established is that there is an analogy between certain *behavioral* features of the two goal-directed processes—for example, it might be shown that both processes are plastic, and that both persist in the face of obstalces provided that these are not too great. In short, whatever may be the merits of the intentional view as an analysis of goal-directed behavior of purposive human beings, it contributes little to the clarification of the concept as it is used in biology.

3

The second account of goal-directed processes I want to examine is sometimes formulated in the language of contemporary information theory, and more specifically in terms of the notion of a "coded program." I will therefore refer to it as the "program view" of goal-directed processes.

As is well known, the materials physically transmitted from parents to progeny in biological inheritance are chiefly the genes located in the chromosomes found in the nuclei of cells. The genes themselves consist of various kinds of very large molecules, among others of nucleic acid molecules called DNA. The DNA molecules have a ladderlike structure, and contain four kinds of nitrogen bases in addition to other chemical groupings. Triplets of these nitrogen bases are sometimes said to be the "letters of the genetic alphabet." Provided that no mutations have occurred in the inherited genes, the sequential order of these "letters" in the long DNA molecules is then described as the "inherited information" (sometimes as the inherited "message" or "instruction") that controls the replication of the molecules, the development of the fertilized eggs, as well as the patterns of activity of the developed organism. Another way of stating this important conclusion of molecular biology is to say that the sequential order of those triplets is the inherited "program" that controls or places limits upon, but without determining exhaustively, the development and the numerous activities of the organism possessing it.

It is this notion of program that has been recently used to explicate the concept of goal-directed behavior; and I will examine the proposal of Professor Ernst Mayr to define it. He first distinguishes between two sorts of "end-directed" processes. Those he describes as "teleomatic" are processes that are said to be "regulated by external forces and conditions," and to achieve their end-state "in a passive, automatic way."[9] The motion of a stone dropped from a tower is an example of a teleomatic process, since the motion is governed by the external gravitational force, and its end-state is reached in an "automatic" way when the stone comes to rest on the ground.

The second sort of "end-directed" processes are called "teleonomic"—the word Mayr adopts to replace the more familiar adjective "teleological," because the latter has been used to cover a miscellaneous assortment of phenomena, and has connotations many of which he wishes to exclude. He defines a "teleonomic" process or behavior as "one which owes its goal-directedness to the operation of a program";[10] and a "program" is tentatively defined as "coded or prearranged information that controls a process (or behavior) leading it toward a given end."[11] Accordingly, a process that has no programmed end cannot properly be designated as teleonomic. However, a program must not only provide "instructions" for achieving a goal; it must also "prescribe" physiological homeostats for dealing with internal and external disruptions of processes leading to the goal.[12] The examples Mayr cites of teleonomic processes in biology include (either explicitly or by implication) the migration of birds, the development of zygotes into adult organisms, and the meiotic process that reduces the number of chromosomes in the sex cells of organisms to the haploid number found in the gametes.

Two additional features of Mayr's account of the program view need to be noted. In the first place, his definition of "program" is admittedly so constructed that no chasm separates ostensibly goal-directed behaviors in organisms from those of man-made machines. For example, he counts as programmed procedures both the repeated tossing of dice loaded so as to favor a given pair of sides turning uppermost, and the behavior of a clock made so as to chime on the hour.[13] In the second place, he counts as programs both the *closed* programs controlling teleonomic processes in organisms that are fully contained in the DNA molecules, and the *open* programs controlling teleonomic behavior, especially in the higher organisms, that permit the incorporation of additional information acquired through learning and other experiences. For example, the *capacity* for song of the white-crowned sparrow, as well as the species-specific *skeleton* of the song, are controlled by the inherited program of these birds; but the population-specific *regional "dialect"* of the song is learned, so that in respect to the distinctive regional features of the song the program is an open one.[14] How various portions of a program come into being—whether transmitted genetically or acquired by learning—has therefore no bearing on the question whether the program leads to a "predictable goal."[15]

As has already been noted, the concept of programmed biological processes is based on some remarkable findings of molecular biology, and it introduces a unifying perspective on important features of goal-directed behavior. However, the relevant question in the present context is: What does the concept of programmed action contribute to the clarification of the notions of being a goal and being goal-directed as these are used in biology?

a) In the first place, it is obvious that in general we do not ascertain whether a process is goal-directed by examining the program that controls it. The coded information that constitutes the program is a complex physico-chemical structure in the DNA molecules located in the genes of the organism under discussion; and at present, at any rate, we do not know just what is the specific physico-chemical structure that corresponds to a given process. We do not know, for example, just what is the sequence of nitrogen bases in the DNA molecules that corresponds to, and is the program for, the meiotic process in cells or the singing of sparrows. Moreover, even if we did know these correspondences, it would be quite difficult to ascertain for a given organism what is contained in its coded program. Accordingly, there must be some ways *other* than by consulting its coded program, of finding out whether a given process is goal-directed, and if it is, just what is its goal. There are indeed such other ways. It is at least a plausible conjecture that an examination of these other ways is more likely to provide effective criteria for identifying goal-directed processes than would an inspection of coded programs.

b) In the second place, the fact that a process is controlled by a program does not suffice to make the process a goal-directed one, unless it is made so by definition. But if such a definitional maneuver is not adopted, a program will correspond to a goal-directed process only if the program is of a *special kind*. For example, the knee-jerk reflex action in normal human beings, which is manifested when the tendon below the kneecap is tapped, is presumably controlled by an inherited program. However, this is not a process that is commonly regarded as goal-directed, chiefly because (as I think) it lacks *persistence*—that is, should some internal disturbance prevent the knee-jerk from taking place, even though the petellar tendon is tapped, the human body is not equipped with any mechanism for making appropriate compensatory responses to those disturbances. It is thus evident that it is possible to state at least some of the conditions a process must satisfy if it is to be a goal-directed one, in terms of the *manifest features* of the process. But no one has yet succeeded in stating, *in terms of the components and structures of DNA molecules,* what requirements a program must satisfy if it is to count as one that controls a goal-directed process. Accordingly, if we do not have some idea of what goal-directed behavior is—an idea that, however vague it may be, is acquired and understood independently of Mayr's formal definition of teleonomic processes—the formal definition will neither convey the idea, nor provide instructions for applying it to biological phenomena, nor clarify the notion of being goal-directed.

c) In the third place, although Mayr's distinction between closed and open programs is an important one, it is debatable whether he is correct in claiming that the origin of a program has no bearing on the question whether the program leads to a "predictable goal." For it is an empirical matter that cannot be settled *a priori,* whether predicting the goal of a process on the assumption that the process is open is as reliable as the prediction of a goal when it is known that the process is closed. I do not know whether a serious study of this question has ever been made. It is my impression, however, that the available evidence favors a negative answer to this question. On the other hand, it is at present difficult if not impossible to ascertain by actually examining a *program,* as distinct from the overt *process* it controls, whether the program is closed or open. The distinction is therefore of no help at present for deciding whether a given process has a predictable goal.

d) Finally, it would be pointless to explicate the notion of being goal-directed if all processes whatever were just of that sort—that is, if processes said to be goal-directed did not differ in some identifiable respect from processes not so characterized. But although Mayr sees this clearly, it is not clear that the distinction he draws between teleomatic and teleonomic processes, with the intent of explicating that difference, attains his objective.

It will be recalled that teleomatic processes are said by Mayr to be "the simple consequences of natural laws" and are "regulated by external forces." However, if the earmark of a teleomatic process is its regulation by "external forces," then the process in which a radioactive substance such as uranium radiates energy is not teleomatic. For the process is not controlled by conditions external to the substance. Moreover, if it can be said that this process is "programmed"—despite the fact that the word would then be used, following Mayr's own practice, in a sense far more inclusive than the one associated with the word in molecular genetics—the process must therefore count as teleonomic. But this is contrary to what I think is Mayr's intent, as well as to what would be generally maintained. One of his examples of a teleomatic process is the behavior of a rock dropped into a well; and one of his examples of a "programmed" (and presumably teleonomic) process is the behavior of a clock built to strike on the hour. Now the rock's behavior is surely teleomatic and not goal-directed, for it arrives at its end-state "automatically." However, since the clock's behavior is also the consequence of relevant laws of nature conjoined with a number of boundary and initial conditions, it is difficult to see why this behavior should not also be described as teleomatic, and as reaching its recurrent end-states automatically. I do not know how to escape the conclusion that the manner in which teleomatic and teleonomic processes are defined does not provide an effective way of distinguishing between processes in biology that are goal-directed from those that are not. In consequence, though the program view notes some important features of goal-directed processes, it is not an adequate explication of the concept.

4

The third account of goal-directed behavior I wish to examine makes use of some of the ideas, though for the most part not the language, of cybernetics, systems thoery, and what is known as the "organismic" standpoint in the philosophy of biology. For reasons that will soon be apparent, I will refer to it as the "system-property" view of goal-directed processes.

Partial anticipations and variant expressions of some aspects of this view can be found in a number of writers; but until the publication at mid-century of a somewhat neglected book by Gerd Sommerhoff, it had no precise systematic formulation.[16] Sommerhoff's principal aim was to state the general conditions a process must satisfy if it is to be classified as goal-directed, irrespective of the particular nature of the systems involved—that is, irrespective of whether the goal is pursued by purposive human agents, by living systems incapable of having intentions, or by inanimate systems. An important and difficult part of this aim was to produce an objective and reasonably precise characterization of what it is that distinguishes processes in biology that supposedly are goal-directed from those which are not. It is easy to see that this complex aim requires that these conditions, as well as the characterization, be formulated in behavioral terms and in a highly abstract manner. In consequence, I can only sketch the central ideas of the system-property view.

Let me first recall two somewhat vaguely described behavioral features that have already been mentioned, concerning which there is good agreement that goal-directed processes have them. One feature is the *plasticity* of such processes—that is, the goal of such processes can generally be reached by the system following alternate paths or starting from different initial positions. The second feature is the *persistence* of such processes—that is, the continued maintenance of the system in its goal-directed behavior, by changes occurring in the system that compensate for any disturbances taking place (provided these are not great) either within or external to the system, disturbances which, were there no compensating changes elsewhere, would prevent the realization of the goal. These features can be regarded as identifying marks for ascertaining whether a process does indeed have a goal, and if so what it is. At all events, it seems to me unlikely that a proposed explication of the concept of being goal-directed would be judged as adequate if it did not incorporate these features into its analyses.

Let us now consider how the system-property view does incorporate them, by analyzing a simple example. In a normal human being the water content of the blood is about 90 percent. This concentration remains fairly steady, despite the constant subjection of the blood to various influences that would alter that percentage were there not compensating changes in the body to prevent it. Call the maintenance of this constant concentration of water in the blood the goal G of the process. The mechanisms that achieve this homeostasis are quite complex; but the general idea of how they work can be readily understood if the simplifying assumption is made that just two sets of organs in the body are involved in that process. One set is the kidneys, which can reduce the concentration of water in the blood; and let K be a variable whose values indicate the quantity of water they remove from the blood. The second set consists of numerous muscles and the skin, all of which can release stored water into the blood-stream; and let M be a variable whose values indicate the amount of water so released. It should be noted that these variables are independent of (or "orthogonal" to) each other, in the sense that within certain limits the value of either variable at a given moment is compatible with *any* value of the other variable at that *same* moment.[17] As will be seen presently, such orthogonality of variables is an important requirement. If now a person drinks much water, his blood does not become diluted, because the kidneys become more active (or are "adapted," in respect to the goal G, to the momentary fluctuations in the blood's water content caused by the drinking), and the value of K increases. On the other hand, if the body loses much water (for example, by sweating), water is released from the muscles and skin into the blood-stream, and M increases.[18]

However, for the process to count as goal-directed on the system-property view, it is not sufficient that on some given occasion (or even on several occasions) the kidneys "just happen" to eliminate excess water from the blood and so "happen" to keep constant the concentration of water in it. An analogous observation has to be made about the muscles and skin. To be goal-directed, the process must satisfy the much stronger requirement that *were* the blood inundated with water to a greater or lesser extent than was actually the case, the activity of the kidneys or of the muscles and skin *would* have been appropriately modified. What this amounts to is that for each member of a sequence of possible values (within certain limits) of the water content of the blood, there is a member of the sequence of possible values of K, such that for each pair of these corresponding values the goal G *would* be achieved—that is, for each such pair, the water content of the blood would be 90 percent. And similarly for the sequence of possible values of the water content of the blood, and the sequence of possible values of the variable M. It is evident that the process is both plastic and persistent; and it could also be shown that the relevant variables are orthogonal. The system is therefore goal-directed. It is also evident that being goal-directed is a property of a *system,* in virtue of the organization of its parts.

Although in this example the goal is the product of a homeostatic mechanism, and only two variables were assumed to be relevant to the realization of the goal, the analysis can easily be generalized to cover other types of goal-directed processes involving any number of variables. But to do so would only complicate the discussion, without adding anything of importance to what has already been said. Moreover, since the system-property view is intended to cover goal-directed behavior whether it is biological or inanimate, it is not relevant in an exposition of the view to discuss either the various particular goals that may be pursued in those processes, or the mechanisms operative in them, or the origins of the mechanisms. Goal-directed processes in living systems are patently programmed, containing "instructions" for the development (among other things) of "feed-back" subsystems; and the origins of the programs are left to be explained by evolutionary theory.

A number of difficulties have allegedly been found in the system-property account of goal-directed systems. But the only really serious difficulty in that view with which I am familiar concerns the requirement that the variables relevent to the realization of the goal of a process must be orthogonal, in the sense already stated; and I want to examine this difficulty. Before I do so, however, I must briefly explain why this requirement is important: in short, it is *this* requirement that serves as a formal criterion for distinguishing processes that are goal-directed from those which are commonly held not to be such.

An example will help to clarify this claim. When a ball at rest inside a hemispherical bowl is displaced from its equilibrium position, restoring forces come into play that in the end bring the ball to rest at its initial position. Is this a goal-directed process, whose goal is the restoration of equilibrium? Were the process so classified, *every* process in which some equilibrium state is restored would also have to be designated as goal-directed; and in consequence, the designation would be applicable to well-nigh all processes, so that the concept of being goal-directed would not be differentiating, and would therefore be superfluous. On purely "intuitive" grounds, however, the answer to the question just raised is negative—an answer which is also in accordance with the orthogonality requirement. For the controlling variables of the ball's motion are *not* independent of each other, since the restoring force is proportional to the magnitude of the displacement force, though oppositely directed.[19] It seems, therefore, that the question whether a process is goal-directed can be decided on the "objective" grounds stated in the requirement, rather than on the basis of "subjective" intuitions that often vary from person to person.

But how are we to understand the requirement that the value of a controlling variable at a given moment is independent of the value of any other controlling variable at the *same* moment? On pain of contradiction, this surely *cannot* mean that the variables must be orthogonal in those very circumstances in which the system *actually is* in the goal-directed state relative to a given goal *G*. As the example of the homeostasis of the water content of the blood makes plain, a system can be in a goal-directed state *only if* there are determinate relations between the relevant variables. It is just because such relations hold that the variables cannot be completely independent of each other.[20] What the requirement *does* mean is that, apart from those situations in which determinate relations hold between the variables because of their role in goal-directed processes, the known (or assumed) "laws of nature" impose no restrictions on the simultaneous values of the variables. For example, when the so-called Watts governor of a steam engine is not hitched up to the engine, any speed of the engine is compatible with any speed of the arms of the governor; for there are no known laws of nature according to which, in the assumed circumstances, the spread of the arms depends on the engine speed.

However, this presupposes that there are uncontroversial criteria for distinguishing laws of nature from laws that hold only for various specialized structures; and in any case, since the answer to the question whether variables are orthogonal depends on what laws of nature are assumed to be known, the notion of being goal-directed is also relative to that assumption. Thus, on the system-property view, the system consisting of an engine with a governor is goal-directed with respect to the goal of a certain rotation speed of the engine's driving wheel. But if the relations holding between the behavior of the governor and the engine speed were included among the laws of nature, that system could no longer be so characterized. Moreover, while the behavior of the simple pendulum is *not* goal-directed relative to the assumptions of Newtonian mechanics and gravitational theory, before Newton's time that behavior might very well have counted as goal-directed.

Is this relativization a fatal flaw in the system-property view? I do not think it is. Although the answer to the question whether a given process is goal-directed is admittedly relative, in most cases it is relative to *factual assumptions* which can, at least in principle, be tested empirically. The answer is therefore not necessarily subjective or unfounded. Indeed, it is frequently possible to decide beyond reasonable doubt, whether the variables relevant to a given goal are orthogonal, at any rate within the limits of certainty required in a given inquiry. For example, the two variables relevant to the analysis of a woodpecker's search for insect grubs on a tree limb are the position of the bird's bill and the position of the insect grubs; and these variables are clearly independent in the required sense. Moreover, it does not seem to me to be a defect in an account of goal-directed behavior that it recognizes, whether explicitly or by implication, that assumptions of goal-directedness may be mistaken and need to be corrected.

5

To provide explanations for the existence of various kinds of organic structures and for the occurrence of different sorts of vital processes is an ongoing task of biology. I want therefore to conclude this lecture with a discussion of some broad issues concerning the nature of explanations that are often proposed for, or in connection with, goal-directed processes in biology.

a) On the assumption that the system-property view is correct, if only in broad outline, the concept of being goal-directed can be explicated without employing in the analysis any specifically biological notions, and in particular without using any expressions that have a teleological connotation. This remark should not be misunderstood. To say that a robin is hunting for worms in order to feed its fledglings is indeed a teleological explanation of the robin's behavior. However, the content of that statement can also be rendered by the assertion that the robin's behavior is goal-directed with respect to the goal of feeding its young; and *this* assertion, since it does not contain such typically teleological expressions as "in order to," neither is a teleological *explanation*, nor is it *formally* a teleological *statement*. Moreover, again on the assumption that the system-property view is sound, the assertion can be explicated in such a way (whether or not the explication counts as a correct translation of the assertion) that terms like "goal" and "goal-directed," which have strong teleological overtones, do not occur in the explication.

This outcome is hardly surprising. For a major objective in developing the system-property view was to analyze the concepts of being a goal and being goal-directed so that the analytic versions of these concepts would contain no undefined teleological notions. But on the hypothesis that this objective has been attained, explanations of goal-directed processes in biology are in principle possible, whose structure is like the structure of explanations in the physical sciences in which teleological notions have no place.

b) In the second place, explanations that have been proposed in connection with goal-directed processes account for the presence of various items in two different ways. One such way is the explanation of how the goal (or outcome) of some goal-directed process or system is realized, in terms of assumed capacities of the system's various organs, the organization of the system's component parts, and a number of laws concerning the effects produced by the activities of those parts. For example, the homeostasis of the blood temperature in normal human beings would be explained in just this way, if the statement that this temperature is approximately constant were shown to be a consequence of assumptions which assert, among other matters, that the body has glands whose activity increases or decreases the body's metabolic rate; that moisture is evaporated in respiration, and sweating; and that there are determinate relations, expressible as general laws, between these activities and changes in the blood's temperature.

Explanations of this sort are often said to be "causal." They resemble in structure, though not in specific content, typical explanations in the physical sciences. As the example just considered suggests, they are like the latter in accounting for the occurrence of some phenomenon by deriving the statement of its occurrence from assumed laws (or general hypotheses) when these are conjoined with statements of relevant initial conditions. Putting all this briefly, one sort of explanation is in terms of antecedent initial conditions and causal laws; and goal-directed processes, among other things, can in principle be explained in this way. Explanations of this type are not distinctive of the life sciences, they are found in all branches of inquiry, and there is nothing teleological about them.

A second sort of explanation *is* characteristic of biology and other sciences that deal with purposive behavior. These explanations do not account for a phenomenon in terms of antecedent conditions and the mechanisms that produce it. On the contrary, they account, or seem to account, for the occurrence of a process or of some other items in terms of certain *effects* these things have on the system of which they are members, or upon some other components of the system. For example, if it is taken for granted that deer pursued by predators often survive by fleeing, explanations of this second sort seem to account for the existence of various items *upon* which the survival is contingent—such as the anatomical features of deer's limbs that make swift flight possible, or the keen sense of hearing of the animal—in terms of what those items *contribute* to the survival of deer. Unlike explanations of the first type, those of this second type are often said to answer the question *why* some process or organic structure exists, or *why* it exists at just the place and time it occupies—in the sense of the particle "why" that requires mention of some purpose in the answer—by stating certain *consequences* of the process or structure. Such explanations have traditionally been called "teleological"; and it is beyond serious doubt that many biologists as well as philosophers construe them in the way just indicated.[21]

Teleological explanations are the main concern of the second lecture, and I am postponing until then further discussion of their structure. However, it has been recently argued by Professor Robert Cummins,[22] that it is a misconception to suppose that the *only* way teleological explanations can be construed is as inferences from effects to causes—that is, as explanations of the *existence* of some entity in terms of certain *effects* the entity has in the system of which it is a component. He rejects this customary interpretation of teleological explanations, and proposes an alternative to it. By way of constructing a bridge between the discussion of goal-directed processes in the present lecture and the issues to be examined in the second one, I will conclude with some comments on the first part of his essay.

Cummins believes that if teleological explanations are interpreted in the customary manner they face insuperable difficulties. What these difficulties are will be ignored for the present, though they will occupy me in the second lecture. Nevertheless, it is because of that belief that he presents a different reading of such explanations. The heart of Cummins' argument is that while he thinks it is "legitimate" to reason from effects to causes—for example, from the evaporation of water to its atomic constitution, or from the homeostasis of blood sugar to the secretion of insulin in the pancreas—the reasoning is said to be "a species of inference to the best explanation." For example, "our best explanation" of the constant water content of blood can be said to "require" kidneys and muscles, so that presumably we can assert the existence of these organs. "But once we see what makes the reasoning legitimate," Cummins continues, "we see immediately that inference *to* an explanation has been mistaken for an explanation itself. Once this becomes clear, it becomes equally clear that [the assumption that to ascribe a goal to a process is to explain the presence of the item characterized as having that goal] has matters reversed: given that [homeostasis of the water content of blood] is occurring in a particular [organism], we may legitimately infer that [kidneys are] present in that [organism] precisely because [kidneys enter] into our best (only) explanation of [that homeostasis]."[23]

Cummins is thus replacing an interpretation of teleological explanations that takes them to be explanations of the *second* type, by an interpretation that construes them to be explanations of the *first* type—though what is *explained* in the former case becomes a factor in what *does* the explaining in the latter. But is Cummins' interpretation of teleological explanations free from difficulties analogous to those he ascribes to the usual one? On the customary reading of a teleological explanation, some effect such as the homeostasis of the blood temperature (call it E) is explained by assuming the operation of some cause such as the activity of various glands (call it C). On the reading Cummins proposes, the effect E is explained by assuming some explanatory hypothesis (call it H) which is said to "require" the cause C (that is, the activity of various glands) that presumably must be supposed to exist if the hypothesis is accepted as correct. In both cases, the existence of some item C is postulated (on Cummins's proposal *via* the hypothesis H), whose activity is assumed to be at least a partial cause of the effect. But if this is so, the difficulties allegedly facing the customary reading of teleological explanations (according to which causes are held to be inferable from their effects) are matched by difficulties no less serious that face Cummins's proposed interpretation (according to which explanatory hypotheses are supposedly inferable from effects deducible from them).

However reasonable Cummins' proposal may be, it does not show that the difficulties associated with the customary reading of teleological explanations can be outflanked by adopting a plausible interpretation for them that is free of comparable problems. A more extended examination of teleological explanations and their difficulties is therefore unavoidable. This is one of the tasks to be attempted in the second lecture.

B. Functional Explanations in Biology

As I mentioned in the first lecture, this second one will deal mainly with teleological explanations in biology. However, since the structure of explanations for goal-directed processes has already been described, the task that remains is to examine explanations in that subclass of teleological explanations which are usually designated as functional. I must therefore make clear what I understand by functional explanations.

Functional *explanations* are most easily described by distinguishing them from functional *statements* on the basis of differences in grammatical structure and with the help of some examples. A functional statement ascribes a *function* to some object or process, as in assertions like "The function of gills in fish is respiration," or "One function of white blood cells (leucocytes) in human bodies is to defend the body against invading microorganisms." On the other hand, a functional explanation does not *explicitly* ascribe a function to anything—indeed, it does not contain the word "function"—but accounts for the presence of some item in a system (or states *why* the item is there) in terms of the contributions the item makes to, or in terms of certain effects the item produces in, the system of which it is a component—as in the assertions "Fish have gills in order to obtain oxygen" or "Human blood contains leucocytes for the sake of defending the body against invading bacteria."

There are a number of divergent analyses of functional explanations, the differences being often attributable to different conceptions of what it is to be a function in biology. In consequence, an examination of proposed accounts of functional explanations is bound to go hand in hand with an assessment of different explications of the notion of function. I want to discuss several types of such explications that have received some attention in the recent literature.

1

The first type is a teleologically neutral definition of the notion of function—I will refer to it as the "neutral" view—such as the one proposed by Professors Walter Bock and Gerd von Wahlert. According to them, the function of a given item in an organism is the set of all the manifest as well as dispositional properties (including the physico-chemical ones) the item exhibits in diverse circumstances, properties that the item possesses in virtue of its components and their arrangement.[1] It is evident that the attribution of a function to an entity, in this sense of "function," like the ascription of mass or velocity to an object in physics, has no teleological connotations. Explanations of the presence of functions so defined will therefore have the same structure as explanations in the physical sciences, and will raise no issues that are distinctive of biological inquiry.

However, Bock and von Wahlert themselves note that their definition of "function" does not express what is perhaps the generally accepted and most widespread sense of the word. In fact, they graft that customary meaning of "function" on the term "biological role," so that in many contexts the word "function" in its familiar meaning can be used interchangeably with the phrase "biological role." For they define the biological role of a faculty in an organism as "the action or use of the faculty by the organism in the course of its life history." For example, the biological role of the large mucus glands of gray jays is said to be the use of those glands "as a glue to cement food particles together into a food bolus which is then stuck to the branches of trees."[2] Accordingly, questions that may be generated in attempts to clarify the nature of explanations of biological roles—for example, under what conditions an item can properly be said to have a biological role, or what is the objective of explanations of biological roles—appear to be quite similar to questions raised in the analysis of functional explanations (in the familiar sense of "function").

2

A second type of explication of the notion of biological function is based on the assumption that the primary meaning of any teleological term is the one it has when used in statements about actions that are directed by purposive agents toward achieving selected ends. Although this view resembles in some respects the account of goal-directed processes that was designated as the intentional view, the view to be examined is sufficiently different from the intentional view to require independent discussion. I will refer to it as the "selective agency" view. According to it, teleological characterizations of nonhuman behavior constitute a "metaphorical extension" of anthropomorphic concepts. In consequence, when the term "function" is employed in contexts in which human intentions are irrelevant (as when the *natural* function of the heartbeat in vertebrates is said to be the circulation of the blood), there must be strong *analogies* with uses of the word in contexts in which some item has been deliberately instituted or selected to behave in some specified manner (as when the governor on a steam engine is said to have the *conscious* function of regulating the speed with which the engine works.) The natural function of an item is therefore not *any* effect of the item's presence; it is that *particular* effect for the production of which the item had been *selected*.

I will waive the question (as not strictly relevant to the present discussion) whether it is in fact the case that the use of teleological language in nonhuman contexts is a metaphorical extension of its alleged primary use in connection with the pursuit of conscious ends. The question that *is* relevant is whether an account of functional explanations in biology is sound, when the account rests squarely on the supposition that there must be strong analogies between teleological characterizations of human and nonhuman behaviors. A number of such analyses have been proposed, but there is time for discussing only one of them.

In a recent book,[3] Professor Larry Wright offers an account of functional explanations that is admittedly based on the assumption that there is such an analogy. He points out that in the case of so-called "conscious functions"—that is, functions assigned to artifacts by their conscious makers—some entity *i* (which may be an object, property, or process) is "introduced" into, or is identified in, a system S, *for the sake of* the effects or consequences *F* that *i* produces in S. Thus, a governor is introduced into, or made part of, an engine in such a way that the spreading and retraction of the arms of the governor affect the speed with which the driving wheel rotates. More generally, the function of an item *i* in system *S* is an effect *F* that *i* produces. However, the function of *i* is not *any* effect; it is *that* effect for the sake of which the item *i* was selected and placed in the position it actually does have. The governor of an engine produces a variety of effects: it adds to the weight of the engine, it reflects light that would otherwise have traveled in some other direction, it makes sounds when it is spinning rapidly, and it regulates the speed of the engine. Not every one of these effects is the function of the governor; its function is *that* effect for the sake of which the governor was constructed and made part of the engine in just the position it actually has.

In consonance with his view on the primary context of teleological language, Wright believes that explanations of natural functions have the same pattern as do explanations of conscious ones. According to him, therefore, to say that the function of the heartbeat in vertebrates is to circulate the blood, is to say two things: first, that the circulation of the blood is an effect or consequence of the heartbeat taking place in the organism; and second, that the heart is present in the animal and engaged in the activity of beating, *just because* it circulates the blood by beating. However, beating hearts do not exist in vertebrates because they were placed there by some *human* agent who selected them for their ability to circulate the blood. If we exclude the possibility of divine intervention, they occupy the place they do in vertebrates because of the operation of *natural selection*.

Wright's analysis of functional explanations therefore yields the following abstract patterns: A functional explanation of the form "The item *i* is in system *S* in order to do *F*" [whose content is the same as that of the functional statement "The function (or a function) of item *i* in system S is *F*"] is equivalent to the conjunction of two statements having the forms: "*F* is a consequence of *i*'s presence in S" and "The item *i* is in *S* just because *F* is a consequence of *i*'s presence in S."

This analysis is said to be adequate for both natural and conscious functions, with the understanding that in the case of natural functions, *natural selection* takes the place of conscious choice.

Wright thus proposes what he takes to be a "causal" analysis of functional explanations. It is said to be a causal analysis because the *causal relation* between the item *i* and its effect *F* allegedly "plays a role" in bringing about the presence of *i* in the system S.[4] However, Wright is very carful to point out that when a function is ascribed to a *particular* item (e.g., to the heartbeat of President Ford's heart at noon on Election Day in 1976), it is not the existence of *this particular item at the stated time* that is explained by the fact tht the presence of item *i* in S on *that* occasion produces *F*. Or stating this point in terms of the example, it is not the existence of President Ford's heart at noon on Election Day that is explained by the fact that the beating of his heart at *that* time produced the circulation of his blood on *that* occasion. It would be patently false if not absurd to assert the contrary. Wright's claim seems to be that the existence of hearts in general (that is, the existence of that *type* of organ) is explained by the fact that organisms with hearts that circulate the blood are more likely to survive and reproduce than organisms with hearts that do not do so—in short, that organisms with beating hearts that circulate the blood have an advantage in the evolutionary process. Or again stating this in terms of the example, on the construal just presented of Wright's claim, President Ford has a beating heart in his body, because his ancestors had beating hearts that circulated their blood—a circumstance that gave them an advantage for survival. The outcome of Wright's analysis is that when we ascribe a function to an item, we are at the same time explaining why that item is present in the system and occupies the place in the system that it actually does.

However, Wright's proposed explication raises some problems. i) His analysis requires us to say that *F* is a function of an item *i* if, and only if, the item has been selected in some way to be present in the organism just because *F* is an effect of *i*'s presence. It therefore follows that *F* can be asserted to be a function of *i*, if and only if it is *known* (or there are good reasons for *believing*) both that *F* is an effect of *i*, and that *i* had been selected to be present in the organism just because *F* is an effect of the item *i*. In fact, however, biologists commonly do state (often on the basis of experimental findings) that a function of some item *i* in organism S is *F*, but without knowing or believing that one *causal determinant* of *i*'s presence in *S* is that *F* is an effect of *i*. For example, when Willian Harvey showed that one function of the heartbeat is to circulate the blood, he was unfamiliar with the etiology of the heart's formation, or with any causal determinants of the heart's presence in vertebrates. Nor is there any evidence to show that when Walter Cannon ascribed to the adrenal medula, on experimental grounds, the function of accelerating the heart's action in circumstances of emergency for the organism, he believed that the gland having this effect is a causal determinant of the gland's presence in the organism. Such examples can be easily multiplied, and are grounds for scepticism concerning the adequacy of Wright's analysis of functional explanations.

ii) But the main reason for doubting the soundness of his analysis is the questionable validity of Wright's central claim that a functionally characterized item is "where it is" *because* the item has that function. In the first place, the claim is mistaken even in the case of *conscious* functions. Consider, for example, the functional statement that the function of the main spring in a watch is to provide power for rotating various cogwheels in the watch. On Wright's analysis, this is equivalent to saying that the spring does have this effect, and also that the spring is where it is because the spring has that effect. However, the second clause of the allegedly equivalent statement is surely an error. The spring was placed where it is by the manufacturer *not* because the spring is able to rotate wheels (as is required by Wright's analysis), but because the manufacturer *knew or believed* that this was so. For springs of the required sort possess that capacity whether or not anyone knows or believes this; and a spring does not appear in a watch simply as a consequence of its possessing that capacity. It is rather the manufacturer's knowledge that a spring does have the capacity which accounts in part for the spring being "where it is."

In the second place, the claim is incorrect when made for natural functions. To be correct, the presence of beating hearts in vertebrates must be accounted for by assuming that vertebrates whose hearts circulate the blood have been "selected" for survival by natural selection *just because* their hearts circulate the blood. However, in the present context the claim can be understood in two ways, each of which requires attention. On one interpretation, the claim is about the causal conditions for the existence of a *particular* heart at a *stated* time in a *given* organism. We do not at present know in sufficient detail just how the heart is formed in the development of a specified organism from a fertilized egg. But we do know that the organism has a heart because the zygote from which it developed was "programmed" to grow one—that is, we know that the zygote has a definite physico-chemical composition such that, in consequence and under normal environmental conditions, numerous (but still largely unknown) physico-chemical processes take place whose outcome is the heart. It is therefore evident that neither conscious nor natural selection plays a role in the genesis of a particular heart. Nor is there any reason for believing that the heart's causal role in circulating the blood needs to be invoked in accounting for the presence of a particular heart in a given body.

On the second interpretation of the claim under discussion (namely, that vertebrates with hearts exist because hearts produce the circulation of blood), the claim is about the causal conditions for the existence of hearts in the vertebrates that have them. Wright believes that the analysis of explanations in this case closely parallels the analysis of explanations in the case of conscious functions. As he puts it, "just as conscious functions provide a consequence-etiology [that is, behavior that occurs because it brings about some specified end] by virtue of conscious selection, natural functions provide the very same sort of etiology as a result of natural selection."[5] The claim is thus made to rest on a reading of the theory of evolution according to which vertebrates with beating hearts exist, bceause only those vertebrates have been selected to survive by natural selection whose beating hearts circulate the blood. Wright therefore concludes that functional explanations, whether the functions are conscious or natural, all have the same structure. Moreover, since in either case consequence-etiologies are asserted, functional explanations are said to be of a distinctive kind, and are not translatable into explanations of the sort customary in the physical sciences.

But is it really the case that natural selection operates, as Wright apparently believes, so as to generate consequence-etiologies—that is, to produce organs of a particular kind just *because* the presence of such organs in the organisms in which they are components gives rise to certain effects? As I see it, to suppose that this is so is to do violence to the currently accepted neo-Darwinian theory of evolution. At the risk of carrying coals to Newcastle, it may be helpful to spell out my reasons for this assertion. According to that theory, *which* heritable traits are possessed by sexually reproducing organisms depends on the genes organisms carry—genes that either are inherited from parent organisms or are mutant forms of inherited genes. Which of its genes an organism transmits to its progeny is determined by random processes that take place during the meiosis and fertilization of the sex cells. It is *not* determined by the *effects* that the genes produce in either the parent or daughter organisms. Moreover, mutation of genes—the ultimate source of evolutionary novelty—also occurs randomly. Genes do not mutate in response to the needs an organism may acquire because of environmental changes; and which genes mutate, as well as what alterations take place in a gene when it mutates, are independent of the effects a gene-mutation may produce in the *next* generation of organisms. Furthermore, natural selection "operates" on individual organisms, not upon the genes they carry; and whether or not an organism survives to reproduce itself does not depend on whether it has traits that would be advantageous to it in some *future* environment. Natural selection is not literally an "agent" that *does* anything. It is a complicated process, in which organisms possessing one assortment of genetic materials may contribute more, in their *current* environment, to the gene-pool of its species than is contributed by other members of the species with different genotypes.

In short, natural selection is "selection" in a Pickwickian sense of the word. There is nothing analogous to "foresight" in its "operation"; it does not account for the occurrence of organisms with novel genotypes; it does not control environmental changes that may affect the chances organisms have of reproducing their kind; and it does not preserve organisms that have traits which are *disadvantageous* to the organisms in their *present* environment, but which may be advantageous to them in *different* environments. The term "natural selection" is thus not a name for some individual object. It is a label for continuing sequences of environmental and genetic changes in which, partly because of the genetically determined traits organisms possess, one group of organisms is more successful, in a given environment, than are other groups of organisms of the species in reproducing their kind, and in contributing to the gene pool of the species. For these reasons, Wright's analysis of functional explanations in biology seems to me to be untenable.

Before leaving the subject of natural selection, it is only fair to add that the supposition that natural selection is more than a negative "sifting" process, and is "creative" in "directing" genetic changes, is endorsed by distinguished students of organic evolution. For example, the late Professor Theodosius Dobzhansky disagreed strongly with those biologists who "doubted that natural selection can be the guiding agent in evolution because selection, allegedly, produces nothing new and merely removes from the population degenerate variants and malformations."[6] He maintained that although the raw materials of evolution are the genotypes that arise by gene and chromosome mutation and recombination, nevertheless "it is selection which gives order and shape to the genetic variability, and directs it into adaptive channels."[7] Indeed, he went on to say that "It is fair to say that selection *produces* new genotypes, even though we know that the immediate causes of the origin of all genotypes are mutation and reproduction. . . . In the long run, selection is the directing agent because it determines which genotypes are available for new mutations to occur in."[8]

A more moderate presentation of the "agent" conception of natural selection is given by Professor Ernst Mayr:

> Evolution is not an all-or-none process. Genetic variation is enormous and does *not* consist merely in the production of a few new types; it "selects" them precisely in the same way in which a breeder "selects" the founder individuals for the next generation of breeding. This is a thoroughy positive process; inferior zygotes are simply lost. We do not hesitate to call a sculptor creative, even though he discards chips of marble. As soon as selection is defined as differential reproduction, its creative aspects become evident. Characters are the developmental products of an intricate interaction of genes, and since it is selection that "supervises" the bringing together of these genes, one is justified in asserting that selection creates superior new gene combinations.[9]

It is little wonder that readers of such passages are persuaded that natural selection is an agent operating in a manner closely similar to the action of a conscious agent, such as the action of an animal breeder. But it is fairly clear that not everything in the quoted passages is intended to be taken literally, as the placing of quotation marks around some of the words in them strongly suggests. It is surely not natural selection that literally *produces* genotypes, for this is done by the mechanisms involved in cell division and reproduction. Nor is it natural selection that does any directing, but it is the environment together with the genotype of an organism which determine whether the organism can survive long enough to reproduce its kind. And it is certainly not the case that natural selection "selects" individuals for survival "precisely in the same way in which a breeder 'selects' the founder indivduals for the next generation of breeding." For a breeder deliberately selects the animals he will mate, on the basis of what he knows of the likelihood that the traits for which he is breeding will appear in the *next* or other *future* generation of the mated animals. But natural selection has no eye to the future; and if any zygotes are eliminated by natural selection, it is because they are not adapted to their *present* environment.

3

I must now turn briefly to an account of the function concept which in one respect is like the view just discussed, but which differs from it in construing functional ascriptions as having a purely methodological or heuristic role. I will refer to it as the "heuristic view" of functional ascriptions.

This view resembles the Kantian interpretation of teleological attributions, but it has contemporary adherents as well, and its distinctive thesis can be stated independently of Kant's elaborate conceptual machinery. Kant was an heir to the thought of Descartes and Newton, and subscribed to the principle that all material processes of nature must be explained by "merely mechanical laws." However, the apparently purposive character of the organization and behavior of living things seemed incapable of being understood in terms of "the mere mechanical faculty of motion";[10] and they had to be viewed *as if* they had been produced by design. Kant therefore formulated a second principle, that *some* events cannot be explained on the basis of purely mechanical laws. But these two principles appear to be incompatible; and according to Kant, they would indeed be contradictory *if* they were assertions about the objective constitution of nature. He resolved the antinomy by construing the principles to be "maxims" or regulative principles for guiding inquiry. According to him, the first principle does not demand that the events of nature be investigated *only* within the framework of mechanical laws. On the contrary, accepting that maxim "does not prevent us, if occasion offers, from following out the second maxim in the case of certain natural forms . . . in order to reflect upon them according to the principle of final causes. . . ." On the other hand, though we may follow the second maxim in dealing with biological phenomena, this does not exclude the possibility that after all living things have been produced in accordance with purely mechanical laws. Indeed, in investigating biological organisms *as parts of nature*, we must go as far as we can in our efforts to understand them in terms of mechanical laws; for unless we do so, "there can be no proper knowledge of nature at all."[11] The conclusion to be drawn from all this is that since we cannot really understand how final causes operate except in the case of our own actions, ascriptions of goals and functions to nonhuman organisms and their parts cannot be taken literally, as objective assertions about nature. They must be construed as statements that have only a heuristic value in guiding inquiry into the mechanisms of living organisms.[12]

Something like Kant's views on this subject is also present in C. D. Broad's definition of teleology:

> Suppose that a system is composed of such parts arranged in such ways as might have been expected *if* it had been constructed by an intelligent being to fulfil a certain purpose which he had in mind. And suppose that, when we investigate the system more carefully under the guidance of this hypothesis, we discover hitherto unnoticed parts or hitherto unnoticed relations between the parts and that these are still found to accord with the hypothesis. Then I should call this system "teleological."[13]

Broad believed that living organisms are teleological systems in the sense of his definition. However, he also maintained that the "intelligent being" needed to design and produce the complex systems that constitute living organisms, would have to possess powers of intellect far beyond anything displayed by minds with which we are familiar. It is therefore unclear just what is assumed in the hypothesis that enters into Broad's definition of teleological systems—that is, it seems impossible for minds like our own to know what *would* be the arrangement of parts of organisms that such a superhuman intelligence instituted to achieve his purposes. It is in consequence also unclear how Broad could have known that living organisms really are teleological systems.

Both Kant and Broad were therefore agnostics about the literal truth of any functional ascription. In both cases, the agnosticism has its source in the assumptions that a process cannot properly be characterized as purposive if it can be explained on the basis of physicochemical laws; and that an effect of an organic process can count as one of its biological functions only if that process was *intended* or *designed* to produce the stated effect. Agnosticism concerning the truth of function ascriptions seems to be the price that must be paid for explicating the notion of biological function in terms of conscious intent. But if the notion of being a function can be explicated, as I believe it can be, with no reference to intentions or choices of conscious organisms, the validity of ascribing a function to an entity can be decided by empirical inquiry, without any need for the elaborate make-believe that seems to be an inseparable accompaniment of the heuristic view.

4

An explication of the notion of biological function that seems far more plausible than those discussed thus far, rests on the assumption that functions contribute to the "welfare" (in a sense to be specified) of either individual organisms, or populations of organisms, or the species to which an organism belongs. I will call it the "welfare" view of biological functions. There are several varieties of this view; and commenting on two of them will enable me to state my own views on the subject more clearly.

a) Perhaps the best known and most carefully articulated critique of functional explanations is that of Professor Carl Hempel. Since his evaluation of them has been highly influential, it is appropriate to examine his analysis once more. His analysis is quie familiar to students of the subject; a brief summary of the salient points in his account will therefore suffice.

According to Hempel, the cognitive content of the functional statement "the function of the heartbeat in vertebrates is the circulation of the blood" is stated more explicitly in the following: "The heartbeat has the effect of circulating the blood, which ensures the satisfaction of certain conditions (e.g., supply of nutriment and removal of wastes) that are necessary for the proper working of the organism."[14] Just what is to be understood by "the proper working" of an organism and by the "certain conditions" that are allegedly required for such working, are important questions; but they can be waived for the present. The basic pattern of a functional *statement* is therefore "The function of item *i* occurring in organism (or system) *S* during period *t* and in environmental setting *C*, is to do *n*"; and the schematic form of the import of such statements is "Item *i* in system *S* during period *t* and in environment *C* has the effects *e* that satisfy the conditions *n* which are necessary for the proper working of S."[15]

On the other hand, on the assumption that a functional *explanation* must account for the presence of item *i* in S—that is, it must explain why it is that during *t* and within *C*, item *i* is present in S—Hempel proposes the following schematic form for the explanatory premises: i) during period *t* and in environment *C*, *S* is in proper working order; ii) if *S* is in proper working order, then condition *n* must be satisfied; and iii) if *i* is present in S, then the effect *e* of *i*'s presence in S satisfies the required condition *n*. (Using Hempel's initial example, the explanatory premises for the presence of a heart in a given human body are as follows: i) during a certain period and under normal environmental conditions, a certain human being is in proper working order; ii) if that person is in proper working order, then various conditions, such as his having nourishment, must be satisfied; and iii) if the person has a beating heart in his body, the circulation of his blood, which is an effect of the heart's beating, satisfies the required conditions.) However, if this is a correct analysis of functional explanations, the assumed premises do *not* entail the desired conclusion, so that the presence of item *i* in organism S is *not* explained.

Hempel points out that the flaw in the argument would be removed if the third premise were replaced by its converse, iii'—that is, by a statement of the form: the condition *n* is satisfied *only if* item *i* is present in S. He believes, however, that in general there is no warrant for doing this. Indeed, he thinks that "it might well be that the occurrence of any one of the number of alternatives would suffice no less than the occurrence of *i* to satisfy *n*."[16] But if this is so, the explanatory premises fail to explain why it is item *i*, rather than one of its possible alternatives, that is present in S. Hempel therefore concludes that while functional characterizations may have considerable heuristic merit, they have little if any explanatory or predictive value.[17] The main burden of Hempel's devastating critique of one proposed account of functional explanations is that the presence of some specified item in an organism, which is to be explained in terms of its function, is in general not a *necessary condition* (or is not known to be a necessary condition) for the performance of that function. Accordingly, the presence of a heart in vertebrates is *not* explained by the fact that it has the effect of circulating the blood, because the heart is not the *sole* thing that can do this; for example, artificial pumps when properly connected with the blood vessels could also perform this function.

It is of course beyond dispute that *if* the structure of functional explanations is as Hempel describes it, these explanations fail to explain their ostensible explananda. However, it is questionable whether functional explanations, at least in biology, do in general have the form he indicates. For example, a convincing case can be made for the claim that in normal human beings—that is, in human bodies having the organs for which they are at present genetically programmed—the heart *is* necessary for circulating blood; for in normal human beings there are in fact no alternative mechanisms for effecting the blood's circulation. For physiologists seeking to explain how the blood is circulated in *normal* human bodies have discovered that human bodies have no organs other than the heart for performing that function. The observation that it *may* be (or actually *is*) physically possible to circulate blood by means of other mechanisms is doubtfully relevant to those investigations of how the blood is circulated in normal human beings, upon which physiologists were once embarked.

The denial of the claim that the heart is necessary for circulating the blood appears to derive part of its plausibility from the unprecise way in which the expression "human body"—and more generally the expression "the system S"—is usually specified. Is the normal human body to be counted as the same system as the body whose blood is being circulated by a mechanical pump? The issue Hempel's analysis raises is one that has long been discussed in connection with the "doctrine of the plurality of causes." For example, since death can result from drowning, from gun wounds, from poisons, and so on, it is sometimes said that death has a plurality of causes. But as has been often noted, the doctrine is plausible in this example only because the *causes* of death have been analyzed more precisely and into a larger number of types than has been done for their *effect*. However, if the state of a body whose death has been caused by drowning is compared with a body whose death was the result of gun wounds, it is clear that each of these causes has its own distinctive effect; and it no longer seems so evident that death has many causes. Something like this point seems to be involved in the denial that the heart is necessary for circulation, except that it is the loose way in which the expression "human body" (or "the system *S*") is used that may underlie the denial.

It must of course be recognized that there are systems in which several items have a common function, so that no particular member of the set of such items is necessary for performing that function. Human beings normally have two ears, neither of which, by itself, is necessary for hearing. Nevertheless, in the normal body it is still necessary that one or the other or both ears be present if the organism is to hear. In such cases, functional explanations account for the presence in the system under discussion of a *set* of items.

It seems to me that Hempel should agree, at least in principle, with this defense of functional explanations containing premises stating that some item in a given system is necessary for the performance of a specified function. Despite his doubts that such premises can be validly asserted, he does assume that there are *necessary conditions* (such as the elimination of wastes) for the proper working of organisms. He must therefore have had in mind organisms which are actually found in nature in determinate environments, and which must satisfy certain conditions if they are to flourish. For if we are free to exercise our imagination and deal with mere possibilities, with no limitations placed on the kinds of organisms that may be considered, organisms can be imagined that produce no waste materials and have, in consequence, no need for eliminating them. However, if necessary conditions can be discovered for the "proper working" of organisms in their natural state, what reasons are there for doubting, on general principle, that certain organs and other parts of organisms may be necessary for the performance of the functions that are associated with those organs and parts? In point of fact, examination of standard treatises on the physiology of the human body shows that the great majority of its organs and parts *are* necessary for the performance of their several functions.

Let me return briefly to Hempel's analysis of functional explanations. As has already been noted, it was not part of the task he set for himself to provide an adequate account of what is to be understood by "the proper working" of an organism, or how the "necessary conditions" for such "proper working" are ascertained. Why then did he place a limitation, involving these notions, on the effects of an item that are to count as the item's function? The main reason for his doing this was to exclude as incorrect function ascriptions which, despite the fact that the alleged functions are the effects of stated items, are strongly counterintuitive. Hempel's example of such a counterintuitive statement is: 'A function of the heartbeat in human beings is to produce heart sounds."

However, the validity of the claim that this statement is a mistaken attribution of function—as all function ascriptions are alleged to be which identify a function of an item with *any* of its effects—depends on whether, *in a given environment,* the production of heart sounds satisfies a necessary condition for the proper working of organisms possessing hearts. It is therefore arguable that while in other environments and other periods heart sounds contributed nothing to the proper working of human beings, this is not true in the present environments of human beings, since heart sounds have a diagnostic value for modern physicians. But be this as it may, the requirement Hempel places on effects if they are to count as functions does enable him to achieve his objective (in principle and at least in part) of excluding as incorrect many counterintuitive attributions of function—provided, of course, that saying an organism to be in proper working order is not just another way of saying that the various parts of the organism are performing their functions. To be sure, without a detailed account of what the proper working of an organism is—an account that is bound to require considerable biological, medical, and perhaps even psychological knowledge—it is difficult to apply Hempel's criterion for distinguishing functions from mere effects. On the other hand, just because he leaves largely unspecified what is to be understood by "the proper working" of organisms, his analysis of functional statements has a generality that many other analyses lack. In consequence, many of these other analyses can be subsumed under, and regarded as special cases of, Hempel's formulation of the structure of functional explanations.

b) I want now to examine the version of the welfare view of biological function proposed by Professor Michael Ruse.[18] He presents his analysis in the context of a critique of some views of my own. An examination of his statement of the issues will enable me to assess both his own analysis and his criticism of mine.

i) In presenting my views on biological functions which Ruse criticizes, I argued that the statement: "The function of chlorophyll in plants is to enable plants to perform photosynthesis," is equivalent to another one that no longer contains any functional terms: "When a plant is provided with water, carbon dioxide, and sunlight, it manufactures starch only if the plant contains chlorophyll." And I added the proviso that functional ascriptions presuppose that, and are appropriate only if, the system under consideration (in the sample it is a plant) is "directly organized" or "goal-directed." I also explained that the term "goal-directed" was to be understood in the sense of the system-property view of the notion of being goal-directed.[19]

Ruse believes that this account of functional statements is "fundamentally misconceived," and states two objections to it. The first is directed against the assumption that the presence of chlorophyll in plants is a necessary condition for the performance of photosynthesis. Neither Ruse nor I think this is a weighty objection, and I have already stated my reasons for believing this. On the other hand, I also think that Ruse may be correct in saying that it is "somewhat unfortunate" I placed so much emphasis on the *necessity* of an item's presence for the performance of a stated function. For the emphasis can be construed as an oversight of cases in which an organism is known to have more than one organ for the performance of a function (as in the case already discussed of organisms having two ears for hearing), as well as an oversight of cases concerning which it may be unknown whether more than one organ is present to perform the function. In such cases, it is not the presence of some *single* item that is explained, but rather the presence of one or more members of a *set* of items. An explanandum of the latter sort is in a sense weaker than an explanandum consisting of a single item; but though it is weaker, it is not necessarily a trivial one.

ii) Ruse's second objection is that the requirement mentioned in my proviso, according to which the use of functional statements presupposes that the systems to which they are applied are goal-directed, is "altogether inappropriate." He offers the following example in support of this judgment. Suppose it were true that long hair on dogs harbors fleas, and also that dogs are goal-directed toward survival. Ruse believes that the conditions stated in my account for something to be a function are satisfied, so that I am committed to saying that a function of long hair on dogs is to harbor fleas. But he thinks this functional statement is strongly counterintuitive, and that no one is likely to assent to it.

On the other hand, he believes that we would be very much inclined to accept that functional statement *if* it were the case that dogs with more fleas receive more fleabites, and that fleabites provide dogs with immunity from a certain parasite whose presence in dogs without fleabites lowers their life-span. The reason why we would be inclined to accept the statement in this case, Ruse thinks, is that on the assumed evidence, harboring fleas contributes significantly to the survival and reproduction of long-haired dogs. In short, on the stated hypothesis, harboring fleas is what biologists call an "adaptation," which confers an adaptive advantage on long-haired dogs. Generalizing from this example, Ruse maintains that to say that the function of item *i* in organism *S* is to do *F*, is to say two things: first, the organism *S* does *F* by using item *i;* and second, *F* is an adaptation—that is, *F* contributes to the survival and reproductive activity of *S*. He therefore concludes that it is a mistake to say, as I did, that a functional statement presupposes that the system about which it is made is goal-directed.

Let me first comment on this second objection. In my opinion, this criticism rests partly on a misunderstanding for which I am largely responsible. I did indeed say that a functional statement "presupposes" that the system under consideration is directively organized or goal-directed.[20] (I did *not* say that a functional statement *implies* that the system is goal-directed, which is the way Ruse states my view. Although the distinction between implying and presupposing is an important one, I do not think that much hangs on this distinction in this portion of Ruse's criticism. I will therefore ignore those parts of his objection that are based on a conflation of the two terms of the distinction.) In any case, what I did say appears to warrant Ruse's claim that to be consistent, I must accept as correct his allegedly counterintuitive statement that the function of long hairs on dogs is to harbor fleas.

However, I also said at other places in my analysis of functional notions that functional statements not only presuppose that the systems under discussion are goal-directed, but also that the function ascribed to an item *contributes* to the realization or maintenance of some goal for which the system is directively organized.[21] I failed to stress this second part of my requirement, and was therefore insufficiently clear in presenting my analysis. But however this may be, I want to make explicit in a semiformal manner what I think is a presupposition in the application of functional statements. A functional statement of the form: a function of item i in system S and enviroment E is F, presupposes (though it may not imply) that S is goal-directed to *some* goal G, to the realization or maintenance of which F contributes. I will call this account as the "goal-supporting" view of biological functions. Since in Ruse's initial example harboring fleas apparently does not contribute to the maintenance of any goal for which dogs are goal-directed, I do not believe that I am committed to holding that the function of long hair on dogs is to harbor fleas. There may be difficulties in my account of functional statements, but the one Ruse mentions does not seem to be one of them.

One difficulty which I do recognize is that *every* effect of an item will have to count as one of its functions, *if* it should turn out that *each* effect contributes to the maintenance of *some goal or other*. Although I have no reason for thinking that this is the actual situation for any organism, I do not know how to eliminate this possibility. However, even if it should be the case that each effect of an item contributes to some goal, it would not follow that the notion of being a function would not be differentiating—that is, it would not follow that every effect of an item would be a function *simpliciter*. For on the goal-supporting view, being a function is relative to *some goal*, but not necessarily relative to the *same* goal.

The goal-supporting view of functions seems to me to be compatible with, but to be more general than, the accounts of both Hempel and Ruse. Ruse appears to deny that this is so. On the strength of the familiar Fregean distinction between meaning and reference, he maintains that the statement that an organism has an adaptive trait *does not imply* that the organism is goal-directed, even if, *as a matter of fact*, all actual organisms were goal-directed for survival. However, since the notion of what is an organism has not been precisely defined in the present discussion (or for that matter in any other discussion), it is endlessly debatable, though perhaps impossible to decide, whether or not there is that implication. In any event, it is not essential that there be that implication for the claim to be correct that the goal-supporting view of functions is more general than those proposed by either Hempel or Ruse. It is sufficient, to establish that claim, that in point of fact all biological organisms are directively organized with respect to *some* goal.

iii) I must comment briefly on Ruse's own thesis that what are called biological functions of various items in organisms are adaptations. It is certainly true that many effects of items that are generally called functions are adaptations. But this does not seem to be so invariably. Whether a designated feature of an organism is an adaptive trait depends on the environment in which the organism lives. In consequence, if the environment is changed, a feature that was adaptive in the earlier environment may not be adaptive in the altered one. For example, the fur of polar bears helps prevent heat loss in the animal, so that in arctic regions possession of heavy fur has an adaptive value for the animal. But what if the environment of polar bears were changed, whether because of long-lasting climatic changes in the polar regions, or because of a migration of polar bears to other climes? In that eventuality, possession of heavy fur may no longer contribute to the survival and reproduction of the bears, although it might still be maintained that one function of the fur is the prevention of heat loss in those animals.

Moreover, it is not incompatible with currently accepted evolutionary theory to suppose that the appearance of an adaptive trait in an organism is strongly coupled with the appearance of another trait which *is* not, or *is not known to be* adaptive, but comes to be designated nonetheless as a function of some item. More generally, biologists frequently succeed in ascertaining effects of some item which they designate as the item's functions, without being at all sure that those effects contribute anything to the survival of organisms possessing the item. For example, certain genes of the yellow onion produce the yellow color of the plant, so that the production of this color is a function of those genes. However, although yellow onions are resistant to a fungus disease while white onions are susceptible to it, the *color* of yellow onions appears to have no adaptive value in itself.[22] This objection to Ruse's thesis is perhaps not a fatal one. But the objection does suggest that making adaptedness the criterion of a trait being a function, is not always congruous with biological practice.

5

In conclusion, I want to state in summary manner the main outcome of these lectures that bears on the nature of functional explanations. In the first place, if the system-property account of goal-directed processes is sound, goal ascriptions can be explicated without employing any teleological notions in the explication; and goal-ascriptions can be explained in a manner that is structurally identical with explanations in the natural sciences. And in the second place, if the goal-supporting view of biological functions is correct, functional statements, as well as the presuppositions of functional ascriptions, can also be rendered without using functional concepts; and functional explanations can be shown to have the same structure as explanations in the physical sciences.

However, one further question must be faced. It has been argued that explanations of both goal and function ascriptions are *structurally* similar to *causal* explanations in the physical sciences. Are the former two kinds of explanation also *causal* accounts, in one case of goal-directed activity, and in the other case of the presence of some item to which a function is attributed? The two cases require separate discussion.

a) It was noted earlier[23] that a statement of the form "The system *S* is goal-directed with respect to the goal *G*" (for example, "The normal human body is goal-directed with respect to the homeostasis of the blood temperature") is explained if it is shown to be a logical consequence of an assumed set of explanatory premises, some of which are held to be laws (for example, premises that include assumptions such as that the human body possesses adrenal glands whose activity affects the body's metabolic rate, that moisture is evaporated when the body sweats, and that these activities produce changes in the temperature of the blood). A number of these premises are causal laws, which state just how the goal *G* is related to various antecedent conditions. Explanations of goal ascriptions are therefore *causal.*

b) Explanations of function ascriptions cannot be characterized in the same way. This will be evident from a consideration of a function ascription having the form: "During a given period *t* and in environment *E*, the function of item *i* in system *S* is to enable the system to do *F*"—for example, "During a period when green plants are provided with water, carbon dioxide, and sunlight, the function of chlorophyll is to enable the plants to perform photosynthesis." The explanatory premises for the assertion having the form "The item *i* occurs in *S* during a given period *t* and circumstances *E*"—for example, "During a stated period and given circumstances, chlorophyll is present in a specified green plant"—are as follows: i) "During a stated period, the system S is in environment E (for example, "During a stated period, a green plant is provided with water, carbon dioxide, and sunlight"); ii) "During that period and in the stated circumstances, the system S does F" (e.g., "During the stated period, and when provided with water, carbon dioxide, and sunlight, the green plant performs photosynthesis"); iii) "If during a given period t the system S is in environment E, then if S performs F the item i is present in S" (e.g., "If during a given period a green plant is provided with water, carbon dioxide, and sunlight, then if the plant performs photosynthesis the plant contains chlorophyll"). It is obvious that "The system S contains the item i during the stated period and in the specified circumstances"—in the example, "Chlorophyll is present in the given green plant"—follows from the premises. The first two premises are instantial statements, and the third is lawlike. However, the performance of F (photosynthesis) is not an *antecedent* condition for the occurrence of the item i (chlorophyll), and so the premise is not a causal law. Accordingly, if the example is representative of explanations of function ascriptions, such explanations are *not* causal—they do not account *causally* for the *presence* of the item to which a function is ascribed.

What then is accomplished by such explanations? They make explicit *one* effect of an item i in system S, as well as that the item must be present in S on the assumption that the item does have that effect. In short, explanations of function ascriptions make evident one role some item plays in a given system. But if this is what such explanations accomplish, would it not be intellectually more profitable, so it might be asked, to discontinue investigations of the *effects* of various items, and replace them by inquiries into the *causal* (or antecedent) conditions for the occurrence of those items? The appropriate answer, so it seems to me, is that inquiries into effects or consequences are as legitimate as inquiries into causes or antecedent conditions; that biologists as well as other students of nature have long been concerned with ascertaining the effects produced by various systems and subsystems; and that a reasonably adequate account of the scientific enterprise must include the examination of both kinds of inquiries.

c) None of these conclusions concerning the character of explanations of goal and function ascriptions shows that the laws and theories of biology are reducible to those of the physical sciences, although if the conclusions really do hold water, they undermine one objection that is sometimes made to the possibility of such a reduction. What I think those conclusions do establish is that teleological concepts and teleological explantions do *not* constitute a species of intellectual constructions that are inherently obscure and should therefore be regarded with suspicion.

Notes

Introduction

1. Max Born, *Physics in My Generation* (London & New York, Pergamon Press, 1956).
2. James B. Conant, *Modern Science and Modern Man* (New York, Columbia University Press, 1952).
3. In his recent book *Progress and Its Problems* (Berkeley: University of California Press, 1977), Professor Larry Laudan denies that science is a "truth-seeking enterprise," so that according to him the rationality of science cannot be usefully defined in terms of the ability of science to attain the truth. As far as I have been able to make out, however, the sole reason he offers for this denial is that "no one has been able to demonstrate that a system like science . . . can be guaranteed to reach the 'Truth'. . . ." (p. 125). But although the fact stated in this reason is not in dispute, the reason is surely irrelevant to the question whether or not science is a truth-*seeking* undertaking. Nor does the admission that there is no demonstrative proof that science will reach the truth, provide any ground for rejecting the hypothesis that on some matters and in some circumstances science does in fact arrive at the truth.

   But however this may be, Laudan proposes an explication of the notion of rationality in terms of scientific progress, an explication that in some respects is similar to the account in the above text. On the general assumption that science seeks to solve experimental as well as "conceptual" problems, he states that the aim of science is to maximize the scope of solved empirical problems and to minimize the scope of anomalous and conceptual ones. The "over-all problem-solving effectiveness" of a theory is then said to be determined by "assessing the number and importance of the empirical problems which the theory resolves and deducting therefrom the number and importance of the anomalous and conceptual problems which the theory generates" (p. 68). When a theory *A* replaces a theory *B*, progress is said to occur if the degree of problem-solving effectiveness of *A* is greater than that of *B*. And finally, the rationality of science is identified as that feature of the enterprise which "consists in making the most progressive theory choices" (p. 6).

   Laudan's explication of rationality resembles the one presented in the text in that in both proposals rationality is associated with the successful achievement of some goal. However, his proposal is faced with a number of serious conceptual problems; and unless they are resolved, his notion of rationality is unusable. For example, how does one assess the number and importance of the empirical problems solved by a theory? Newtonian gravitational theory makes it possible to compute the orbits of each of the nine planets. Has the theory solved nine problems or just one? And what standard is to be employed in deciding whether or not the problem of calculating planetary orbits is more important than the problem of accounting for the earth's tides? Again, is it the sheer *number* of solved problems that counts, or is their *variety* also of moment? And how is one to decide between those who (like Niels Bohr) think that the Copenhagen interpretation of quantum mechanics raises no unanswered conceptual problems, and those who (like Einstein) believe the contrary? Laudan's book provides no answers to such questions, nor does it contain any hints as to how the answers are to be found.

4. They are the sole fruits of a long since abandoned plan to write a comprehensive history of changes during the 19th century in methodological ideas employed in various branches of inquiry—in the natural, psychological, and social sciences, but also in a number of humanistic disciplines (such as history, legal scholarship, and hermeneutics).
5. See, for example, Barry Barnes, *Scientific Knowledge and Sociological Theory* (London and Boston: Routledge & Kegan Paul, 1974). In his rejoinder to the claim that scientists do not select theories for ideological reasons, but "test them in good faith without prior commitment," Barnes writes: "We must again ask where the initial set of theories comes from. If, as presumably it must be, it is drawn from, or inspired by the scientists' general cultural resources, then the final, rationally chosen, theory may be partially determined by social factors" (pp. 12–13). It seems reasonable to infer from this that scientists cannot be the source of genuinely novel ideas which are not "drawn" from the ambient society, and that if such novelties apparently do occur their ultimate source *must* be the scientists' "culture." Just why this is so is not explained in Barnes's book.
6. "Weimar Culture, Causality, and Quantum Theory, 1918–1927," in *Historical Studies in the Physical Sciences*, ed. Russell McCormmach, Third Annual Volume (Philadelphia: University of Pennsylvania Press, 1971).
7. Ibid., p. 110.

2. Theory and Observation

1. Albert Einstein, *Ideas and Opinions* (New York: Crown Publishers, 1954), pp. 271–72.
2. *Philosophical Transactions of the Royal Society* 6 (1671/72), reprinted in *Isaac Newton's Papers and Letters on Natural Philosophy*, ed. I. B. Cohen (Cambridge, Mass.: Harvard University Press, 1958), pp. 47–48.
3. Ibid., p. 53.
4. There are good reasons for believing that even Newton's theoretical notion of a light ray is independent of various further assumptions that were available to him concerning the nature of light—e.g., the assumption that light is corpuscular, that it is undulatory, or that it travels through an all-pervading optical medium. Cf. Robert

Palter, "Newton and the Inductive Method," *The Texas Quarterly* 10, no. 3 (1967), p. 168.

5. For example, Paul Feyerabend, "Problems of Empiricism," in *Beyond the Edge of Certainty,* ed. Robert G. Colodny (Englewood Cliffs, N.J.: Prentice-Hall, 1965); N. R. Hanson, *Patterns of Discovery* (Cambridge: Cambridge University Press, 1958); Mary Hesse, "Theory and Observation: Is There an Independent Observation Language?," in *The Nature and Function of Scientific Theories,* ed. Robert G. Colodny (Pittsburgh, Pa.: University of Pittsburgh Press, 1970); Thomas S. Kuhn, *The Structure of Scientific Revolutions* (Chicago: University of Chicago Press, 1962); Stephen Toulmin, *Foresight and Understanding* (Bloomington: University of Indiana Press, 1961). In what follows, I have made extensive use of Mary Hesse's essay, and I am indebted to her for formulations of some of the issues.
6. Paul Feyerabend, p. 213.
7. Mary Hesse, p. 46.
8. See, for example, Paul Feyerabend, p. 214.
9. Mary Hesse, p. 62.

4. The Quest for Uncertainty

1. Charles Sanders Peirce, *The Philosophy of Peirce,* ed. Justus Buchler (New York: Harcourt Brace, 1940), p. 53.
2. Ibid., p. 18.
3. Ibid., pp. 55–56.
4. Karl Popper, *Conjectures and Refutations* (New York: Basic Books, 1963), p. 27.
5. Ibid., p. 229.
6. Karl Popper, *The Logic of Scientific Discovery* (New York: Basic Books, 1959), pp. 278–79.
7. Popper, *Conjectures,* pp. 387–88.
8. Ibid., p. 118.
9. Popper, *Logic,* pp. 105, 109.
10. Popper, *Conjectures,* p. 238; my italics.
11. Popper, *Logic,* p. 104.
12. Ibid., p. 109.
13. Ibid., p. 105.
14. Ibid., p. 394.
15. Ibid., p. 415.
16. Ibid.
17. Ibid., pp. 418–19.
18. Popper, *Conjectures,* p. 217.
19. Ibid., p. 241.
20. Ibid., pp. 243–44.
21. Ibid., p. 222.

22. Ibid., p. 248.
23. Ibid., p. 215.
24. Popper, *Logic*, p. 86.
25. Paul Feyerabend, "Explanation, Reduction, and Empiricism," in *Minnesota Studies in the Philosophy of Science*, Vol. III, ed. H. Feigl and G. Maxwell (Minneapolis: University of Minnesota Press, 1962), pp. 36–37.
26. Max Planck, *Treatise on Thermodynamics*, 3rd ed. (New York: Dover Publications, Inc., 1926), p. 97*n*.
27. George Santayana, *Reason in Science* (New York: Scribner's, 1905–6), p. 24.

5. Philosophical Depreciations of Scientific Method

1. Paul Feyerabend, *Against Method* (Atlantic Highland: Humanities Press, 1975), p. 23.
2. Ibid., p. 26.
3. Ibid., p. 32.
4. Paul Feyerabend, "Problems of Empiricism," in *Beyond the Edge of Certainty*, ed. Robert G. Colodny (Englewood Cliffs, N.J.: Prentice-Hall, 1965), p. 180.
5. Ibid., p. 214.
6. Paul Feyerabend, *Against Method*, p. 285.

6. Issues in the Logic of Reductive Explanations

1. Paul Feyerabend, "Problems of Empiricism," in R. G. Colodny, ed., *Beyond the Edge of Certainty* (Englewood Cliffs: Prentice-Hall, Inc., 1965), p. 180.
2. Paul Feyerabend, "Reply to Criticism," *Boston Studies in the Philosophy of Science* 2 (1962), p. 252.
3. Paul Feyerabend, "Explanation, Reduction, and Empiricism," *Minnesota Studies in the Philosophy of Science* 3 (1962), p. 76.
4. Feyerabend, "Problems of Empiricism," p. 175.
5. Ibid., p. 213.
6. Ibid., p. 216.
7. Ibid.

8. Ibid., p. 175.
9. Ibid., p. 227.
10. *Boston Studies in the Philosophy of Science* p. 231; cf. also Feyerabend, "On the 'Meaning' of Scientific Terms," *Journal of Philosophy* 62 (1965), p. 271.
11. Feyerabend, "Explanation, Reduction and Empiricism," pp. 28–9, 59.
12. Many of them are noted by Dudley Shapere in his "Meaning and Scientific Change," in R. G. Colodny, ed., *Mind and Cosmos* (Pittsburgh: University of Pittsburgh Press, 1966).
13. Feyerabend, "Problems of Empiricism," p. 21; and Feyerabend, "Explanation, Reduction, and Empiricism," p. 24.
14. Feyerabend, "On the 'Meaning' of Scientific Terms," *Journal of Philosophy* 62 (1965), p. 268.
15. "The Relation of the Physical Sciences to Biology," in Bernard H. Baumrin, ed., *Delaware Seminar in the Philosophy of Science* (New York: John Wiley Sons, 1963).
16. Ibid., p. 241.
17. Ibid., p. 242.
18. Ibid., p. 243.
19. Ibid.
20. Ibid., p. 246.
21. Ibid., p. 247.

7. Carnap's Theory of Induction

1. Rudolf Carnap, *Logical Foundations of Probability* (Chicago: University of Chicago Press, 1950), p. 218. This work will be cited in the sequel as *LFP*. Carnap's *The Continuum of Inductive Methods* (Chicago: University of Chicago Press, 1952), will be cited as *CIM*.
2. Carnap's attempt to develop a purely comparative logic of induction, in which only relations of order are assumed between degrees of confirmation but no numerical measures are assigned to them, has not been entirely successful thus far. Carnap has himself recognized that the comparative logic presented in Ch. 7 of *LFP* leads to counterintuitive results—cf. his article "On the Comparative Concept of Confirmation," *British Journal for the Philosophy of Science* 3 (1952–53). I shall therefore omit in this essay all discussion of Carnap's comparative inductive logic.
3. *LFP*, p. 62.
4. *LFP*, p. 75.
5. Rudolf Carnap, "On the Application of Introductive Logic," *Philosophy and Phenomenological research* 7 (1947), p. 137.
6. *LFP*, p. 571 f.
7. *LFP*, p. 574 f.
8. *LFP*, p. 211.
9. *LFP*, p. 212 f.

10. Thus, if we assume that only gravitational forces are present, we can deduce from Newtonian theory certain conclusions about the orbit of a given body. But if there are also magnetic forces in operation which enter into the determination of the orbit and which we have unwittingly ignored, our original conclusions are clearly wrong.
11. *LFP*, p. 571.
12. *CIM*, p. 24.
13. *LFP*, p. 168.
14. Ibid., p. 168.
15. Ibid., p. 170.
16. Ibid., p. 171.
17. *CIM*, p. 25.
18. Ibid., p. 53.
19. Ibid., p. 54 f.
20. Ibid., p. 71.
21. *LFP*, p. 178.
22. Ibid., p. 179.
23. Ibid., p. 180.
24. Ibid., p. 181.
25. I suspect, however, that though a formally valid proof may be given, the proof requires premises which are question-begging, in the sense that the assumption of the world's uniformity is presumably built in antecedently into the c-function used for determining the probability_1 of that assumption.

8. "Impossible Numbers": A Chapter in the History of Modern Logic

1. No explicit mention of Boole's mathematical background is found in the historical accounts given by C. I. Lewis, H. Scholz, or John Venn. No more than a suggestion that there was such a background is given either by L. Liard or by J. Jorgenson; the source of the latter's information seems to be the good but incomplete accounts by P. E. B. Jourdain in the *Quarterly Journal of Pure and Applied Mathematics* (1910). F. Enriques' brief history is characteristically more complete on this point, and G. Vailati's *Scritti* contains good hints as to sources.
2. Published in 1847, the same year as A. DeMorgan's *Formal Logic*.
3. Euler, *Vollständige Anleitung zur niedern u. höhern Algebra* (Berlin, 1796), I, 4. This treatise was first published in 1770.
4. Quoted in Philip Kelland and P. G. Tait, *Introduction to Quaternions* (London, Macmillan & Co., 1904), p. 2.
5. Euler, p. 10.
6. Euler declared: "Weil nun alle mögliche Zahlen, die man sich nur immer vorstellen mag, entweder grösser oder kleiner als 0, oder 0 selbst sind; so ist klar, dass die Wurzel von negativen Zahlen nichts einmal unter die möglichen Zahlen gerechnet

werden kann. Folglich muss man behaupten, dass sie unmögliche Zahlen sind. Und dieser Umstand leitet auf dem Begriff von solchen Zahlen, welche ihrer Natur nach unmöglich sind, und gewöhnlich imaginäre oder eingebildete Zahlen genannt werden, weil sie bloss in der Einbildung statt finden" (*Vollständige,* 71–72).

7. Royal Astronomical Society, *Memoirs* 12 (1842), p. 459.
8. *Life and Writings of Robert Simson* (London, 1812), p. 67.
9. Cf. A. DeMorgan, *Budget of Paradoxes* (Open Court Ed., 1915), I, 203.
10. William Frend, *Principles of Algebra* (London, 1796), pp. x–xi.
11. *Principles of Algebra* (London, 1799), Part Second, pp. ix–x.
12. *Treatise of Algebra,* 1685, Ch. 66, 67.
13. John Playfair, "On the Arithmetic of Impossible Equations," *Philosophical Transactions of the Royal Society* 68 (1778), p. 318.
14. He had declared that the name of reasoning cannot be given to the process of calculating with "impossible numbers."
15. Playfair, p. 321.
16. Ibid., p. 326.
17. Ibid., p. 342.
18. *Edinburgh Review* 12 (1808), p. 306 ff. He continued, with British obstinacy: "The author [i.e., Buée] has, by certain metaphysical subtleties, set himself above the plain dictates of elementary science. We do not, indeed, very clearly comprehend many of the subtleties, and we do not much lament that we feel an incapacity of doing so. It seems very useful, on some occasions, to have one's head fortified with a decent degree of obtuseness to prevent the influx of false refinements, which, when suffered to intrude themselves into the mind, are very apt to dispossess the lawful inhabitants" (p. 312).
19. G. Loria, "L'Enigme des nombres imaginaires à travers les siècles," *Scientia* 21, Supp., p. 43 ff.
20. Bueé, "Memoire sur les quantités imaginaires," *Philosophical Transactions of the Royal Society* 96 (1806), p. 27.
21. On the question of priority between Buée and Argand, there is an interesting correspondence between DeMorgan and Sir Wm. R. Hamilton, *Life of Sr. Wm. R. Hamilton* by R. P. Graves (Dublin, 1882), III, pp. 316, 437–41.

    In Gergonne's *Annales de mathématiques* 5 (1813), Français admitted he obtained the ideas for his own paper on imaginaries from Legendre, who in turn learned them from Argand. Even at this late date it was possible for Servois to object that the line represented by $a\sqrt{-1}$ *cannot be a mean proportional to the lines* $+a$ and $-a$. (Cf. p. 228.) Gergonne replied in effect that Servois forgets that $a\sqrt{-1}$ does not simply represent a quantity but a direction as well; he thus showed he was aware algebraic expressions need not be taken as necessarily representing simple quantities. Gergonne's own paper on negative quantities is a very intelligent discussion of the subject, and makes clear the conventional element in the interpretation of the algebraic signs + and −.
22. C. V. Mourey, *La Vraie Theorié des quantités negatives et des quantite9s prétendues imaginaires* (Paris, 1828). Mourey discusses the view, familiar by then, that a "pure" algebra is not concerned with the interpretation of its symbols (pp. v–vii).
23. "Consideration of the Objections Raised against the Geometrical Representation of the Square Roots of Negative Qualities," *Philosophical Transactions of the Royal Society* 119 (1829), p. 244.
24. Ibid., p. 250.
25. The reader will not be surprised that a nominalistic bias characterized many

members of this group. I do not believe, however, that such a bias is really consistent with the analysis of algebra which this school has contributed.

26. "On the Necessary Truth of Certain Conclusions Obtained by means of Imaginary Quantities," *Philosophical Transactions of the Royal Society* 91 (1801), p. 90. Woodhouse's implicit pragmatism comes out clearly in a letter to Baron Maseras: "Whether or not I have found a logic, by the rules of which operations with imaginary quantities are conducted, is not now the question; but surely this is evident, that, since they lead to right conclusions, *they must have a logic.*" Royal Astronomical Society, Memoirs 12 (1842), p. 462.
27. Ibid., p. 107. Italics not in text.
28. Woodhouse shows clearly that the so-called "demonstrations" of the properties of negative and imaginary numbers (e.g., minus times minus equals plus), so common even in recent elementary texts, are fallacious. He takes special pains to point out that the equivalence between

$$(a + b\sqrt{-1}) \times (c + d\sqrt{-1})$$

and

$$(ac + ad\sqrt{-1} + cb\sqrt{-1} - bd)$$

must be *assumed,* and cannot be *proved* by the principles of arithmetic. Ibid., p. 93. Cf. also his important *Principles of Analytic Calculation* (1803), p. 8.

29. "On the Independence of the Analytical and Geometrical Methods of Investigation," *Philosophical Transactions of the Royal Society* 92 (1802), p. 87.
30. *A Treatise on Algebra* (1830); "Report on the Recent Progress and Present State of Certain Branches of Analysis," *Report of British Association for the Advancement of Science* (1833); and *Treatise on Algebra,* Vol. I (1842) and Vol. II (1845).
31. *Report of the British Association for the Advancement of Science,* Vol. III, p. 186. Peacock goes on: "In the speculative sciences, we merely regard the results of science itself, and the logical accuracy of the reasoning by which they are deduced from assumed first principles: and all our conclusions possess a necessary existence, without seeking either for their strict or for their approximate interpretations in the nature of things; in the physical sciences, we found our reasonings equally upon assumed first principles, and we equally seek for logical accuracy in the deduction of our conclusions from them; but both in the principles themselves and in the conclusions from them, we look to the external world as furnishing by interpretation corresponding principles and corresponding conclusions; and the physical sciences become more or less adapted to the application of mathematics, in proportion to the extent to which our first principles can be made to approach to the most simple and general facts or principles which are discoverable in those sciences by observation or experiment. . . . It is true that there exists a connection between physical and speculative geometry, as well as between physical and speculative mechanics; and if in speculative geometry we regarded the actual construction and mensuration of the figures and solids in physical geometry alone, the transition from one science to the other being made by interpretation, then speculative geometry and speculative mechanics must be regarded as sciences which were similar in their character, though different in their objects: but we cultivate speculative geometry without any such exclusive reference to physical geometry, as an instrument of investigation more or less applicable, by means of interpretation, to all sciences which are reducible to measure, and whose abstract conclusions, in whatever manner *suggested* or derived, possess a great practical value altogether apart from their application to practical geometry" (pp. 187–88).

32. Ibid., p. 189. The second edition of Peacock's *Treatise on Algebra* consequently appeared in two volumes, the first devoted exclusively to Arithmetical, the second to Symbolical, Algebra. Cf. II, p. 1 ff.
33. Ibid., p. 194. "Inasmuch as in many cases, the operations of Symbolical Algebra required to be performed are impossible, in their ordinary sense, it follows that the meaning of the operations performed, as well as of the results obtained under such circumstances, must be derived from the assumed rules, and not from their definitions or assumed meanings, as in Arithmetical Algebra" (*Treatise on Algebra,* II, p. 7).
34. Ibid., p. 197. He adds in a note: "To *define,* is to assign beforehand the meaning or conditions of a term or operation; to *interpret,* is to determine the meaning of a term or operation conformably to definitions or to conditions previously given or assigned. It is for this reason, that we *define* operations in arithmetical algebra conformably to their popular meaning, and we *interpret* them in symbolical algebra conformably to the symbolical conditions to which they are subject." Cf. also *Treatise,* II, pp. 448–49.
35. "The rules of symbolical combination which are thus assumed have been *suggested only* by the corresponding rules in arithmetical algebra. They cannot be said to be *founded* upon them, for they are not *deducible* from them; for though the operations of addition and subtraction, in their arithmetical sense, are applicable to all quantities of the same kind, yet they necessarily require a different meaning when appliecd to quantities which are different in their nature, whether that difference consists in the kind of quality expressed by the unaffected symbols, or in the different signs of affection of symbols denoting the same quantity; neither does it necessarily follow that in such cases there *exists* any interpretation which can be given of the operations which is competent to satisfy the required symbolical conditions" (Ibid., p. 198). The heuristic principle, of taking the rules of operation of arithmetic as suggestions for developing an unrestricted symbolical algebra, was baptized by Peacock as the "principle of the permanence of equivalent forms." His formal statement of it is: "Whatever form is algebraically equivalent to another when expressed in general symbols, must continue to be equivalent whatever those symbols denote. Whatever equivalent form is discoverable in arithmetical algebra considered as the science of suggestion, when the symbols are general in their form, though specific in their value, will continue to be an equivalent form when the symbols are general in their nature as well as in their form" (p. 199). Cf. *Treatise,* II, p. 59. It must be clearly understood that this principle expresses a *resolution* for the construction of symbolical systems, and cannot be used, as Peacock was aware, to demonstrate the rules of combination in the new science. It is possible to construct algebras in which all the rules of common arithmetic do not hold. Cf. Peacock's discussion of multiplication of common fractions, *Treatise,* I, pp. 74–75, or his discussion of the algebraic rules for signs, II, p. 18.
36. *Report,* p. 198. *Treatise,* II, p. 449.
37. *Report,* p. 200. *Treatise,* I, p. vii.
38. E.g., in discussing the imaginaries he says: "The capacity possessed by the signs of affection involving $\sqrt{-1}$ of admitting geometrical or other interpretations, though it adds greatly to our power of bringing geometry and other sciences under the dominion of algebra, does not in any respect affect the general theory of their introduction: for, in the first place, it is not an essential or necessary property of such signs; and in the second place, it in no way affects the form or equivalence of symbolic results, though it does affect both the extent and the mode of their application" (*Report,* pp. 229–31).

39. It is simply an error to assert that until Grassmann's work appeared in Germany, no study was devoted to the mathematical properties of operations, even if Hamilton is made an exception to this dictum. Cf. G. Stammler, *Der Zahlbegriff seit Gauss* (Halle: Max Niemeyer Verlag, 1926), p. 52.
40. "On the Real Nature of Symbolical Algebra," *Edinburgh Philosophical Transactions* 14, Part I (1838), p. 208.
41. DeMorgan, *Elements of Algebra,* 2d ed. (London: Taylor & Walton, 1837), pp. 54–59.
42. "On the Foundations of Algebra," *Transactions of the Cambridge Philosophical Society* 7 (1839), p. 173. DeMorgan continues: "It is desirable that the word *definition* should not enter in two distinct senses, and I should propose to retain it as used in the *art* of algebra, applying the terms *explanation* and *interpretation* to denote the preparatory and terminal processes of *science.* Thus a symbol is *defined* when such rules are laid down for its use as will enable us to accept or reject any proposed transformation of it, or by means of it. A simple symbol is *explained* when such a meaning is given to it as will enable us to accept or reject the application of its definition, as a consequence of that meaning."
43. Ibid., p. 287.
44. *Trigonometry and Double Alegebra* (London: Taylor, Walton & Haberly, 1849), pp. 92–93. Later on, apropos of the signs for addition and subtraction introduced in symbolical algebra, he says: "If any one were to assert that + and − might mean reward and punishment, and *A, B, C,* etc. might stand for virtues and vices, the reader might believe him, or contradict him, as he pleases—but not out of *this* chapter" (p. 101).
45. It is perhaps unnecessary to remind the reader that it is not the *Scotch* Hamilton to whom reference is made here. The two have been confused even by their contemporaries. Corresponding with the *Irish* Hamilton anent his well-known controversy with the Scotch Hamilton, DeMorgan wrote: "The party making the charge [plagiarism] is your namesake, Sir Wm. Hamilton of Edinburgh. If I cannot drive him to press in a week or two . . . I must publish myself; so that all will soon be out. In the meanwhile I send you this information that you may not stare if anybody tells you that *you* are charging me with stealing logic from you. . . . I was talking to a friend on this matter the other day, and I said to him, 'You know Sir W.H. is no mathematician—in fact he is an opponent of mathematics.' I saw my friend's eyes open very wide, and he looked to see if I were gone mad—I had forgotten to say 'of Edinburgh.'" *Life of Sir Wm Rowan Hamilton,* by Robert P. Graves (Dublin, 1882–89), III, p. 266.
46. Ibid., Vol. II, Ch. 17, 18. Cf. p. 141.
47. *Hermathena* 3 (1879). Letter of March 4, 1835.
48. Ibid., Letter of March 13, 1835.
49. *Life,* II, p. 343.
50. "The Theory of Conjugate Functions, or Algebraic Couples, with a Preliminary Essay on Algebra as the Science of Pure Time," Royal Irish Academy, *Transactions* 17 (1837). He regarded the conclusion that "pure Time" was essential to algebra as an induction from the history of mathematics. "The history of Algebraical science shows that the most remarkable discoveries in it have been made, either expressly through the medium of the notion of *Time,* or through the closely connected (and in some sort coincident) notion of *Continuous Progression.* . . . Lagrange, in the Philologic spirit, sought to reduce the Theory of Fluxions to a system of operations upon symbols, analogous to the earliest symbolical operations of Algebra, and professed to

reject the notion of time as foreign to such a system; yet admitted that fluxions might be considered only as the velocities with which magnitudes vary. . . . Lagrange considered Algebra to be the *Science of Functions,* and it is not easy to conceive a clearer or juster idea of *Function* in this Science, than by regarding its essence as consisting in a Law connecting *Change with Change.* But where *Change* and *Progression* are, there is TIME" (p. 295).

51. Hamilton gives a brief account of these investigations in the Preface to his *Lectures on Quaternions* (Dublin: Hodges and Smith, 1853), p. 10.
52. Ibid., Preface, pp. 18–61.
53. Ibid., p. 15. Cf. also his comments on Ohm, a member of the symbolic school, *Life,* II, p. 417: "*Ordinals* seem to me to have, in thought, priority over *cardinal* numbers: 'one,' 'two,' 'three,' mean *originally,* 'first,' 'second,' 'third'; they are names rather of the counted things, than of the groups containing them; and thus even in answering the questions:—How many? we undoubtedly bring in the question, and its answer: Which in a progression?"
54. *Life,* II, p. 522.
55. Ibid., p. 528. He adds: "When I first read that work, . . . it seemed to me, I own, that the author designed to reduce algebra to a mere system of symbols, and *nothing more;* an affair of pothooks and hangers, of black strokes upon white paper, to be made according to a fixed but arbitrary set of rules: and I refused to give the high name of *Science* to the results of such a system; as I should, even now, think it a stretching of courtesy, to speak of chess as a 'science,' though it may well be called a 'scientific game.' "
56. Ibid., p. 575.
57. Boole, *Differential Equations,* 2d ed. (Cambridge, London, 1865), p. 398.
58. Cf. the extract from an undated manuscript of Boole, quoted by Jourdain, *Quarterly Journal of Pure and Applied Mathematics* 41, p. 346. Also, Boole, *Mathematical Analysis of Logic* (London, 1847), pp. 3–6.
59. R. Harley, in a brief biographical account of Boole, in *British Quarterly Review* 64, p. 157.
60. Mary Everest Boole, *Boole's Psychology, as a Factor in Education* (Colchester, 1902), pp. 19 ff. Florence Daniel, *A Teacher of Brain Liberation* (London: C. W. Daniel Co., 1923), p. 10.
61. *Laws of Thought* (London, 1854), pp. 6–7.
62. For example, swerving away from his earlier Platonic theory of logic and mathematics, Russell now declares that propositions of logic and mathematics "are really concerned with symbols. We can know their truth and falsehood without studying the outside world, because they are only concerned with symbolic manipulations." *Analysis of Matter* (London, New York: Kegan Paul, Trench, Toubner & Co.; Harcourt, Brace & Co., 1927), p. 171. And a recent handbook summarizes Hilbert's views as follows: "Pure logic and mathematics are calculi in which the essential thing is the derivation of a set of expressions or formulas from certain initial formulas by means of fixed rules of operation." W. Dubislav, *Philosophie der Mathematik in der Gegenwart* (Berlin: Junker and Dünnhaupt Verlag, 1932), p. 47.

   It need scarcely be said that this approach to logic and mathematics must be supplemented by a careful examination of the different senses in which the word "sign" or "symbol" is employed. It is patent, for example, that one must distinguish between the *type* or *kind* of a sign, and its *hic et nunc* spatio-temporal exemplification. Thus a formula is derivable from a set of initial formulae, not because of the *hic et nunc* distribution of marks on a piece of paper, but because of the *type* of such distribution. Peirce seems to be the only one who has taken the trouble to make the requisite analysis of signs. Hilbert's distinction between "mathematics" (or the manipulation of marks), and "meta-mathematics" (or the principles governing and limiting such manipulation), is also relevant in this connection.

9. The Formation of Modern Conceptions of Formal Logic in the Development of Geometry

1. This doctrine is still enshrined in the definitions supplied by both the Century and Oxford Dictionaries. According to the former, geometry is that branch of mathematics which deduces the properties of figures in space from their defining conditions, by means of assumed properties of space; according to the latter, it is the science which investigates the properties and relations of magnitudes in space.

Barlow's *Dictionary* expresses the outlook at the beginning of the last century: "Geometry, according to the present acceptation of the term, may be defined as the science of extension, or of magnitudes, considered simply, generally, or abstractedly, or rather, it is that science which treats of the relative magnitude of extended bodies." P. Barlow, *A New Mathematical and Philosophical Dictionary* (London: G. & S. Robinson, 1814).

Newton's dictum, stated in the preface to his *Principia,* that geometry is "that part of the universal mechanics which accurately proposes and demonstrates the art of measuring," was much too prosaic for many of his contemporaries and of posterity. To Leibniz, for example, arithmetic and geometry are both "innate, and are in us virtually, so that we can find them there if we consider attentively and set in order what we already have in the mind, without making use of any truth learned through experience or through the tradition of another." G. W. Leibniz, *New Essays Concerning Human Understanding* (Chicago: The Open Court Publishing Company, 1916), p. 78. However, Leibniz insisted that geometry is the science of extension. Ibid., p. 700. Euler's views may be taken as representative of the most matured ideas on the subject in the eighteenth century: "Extension is the proper object of geometry, which considers bodies only in so far as they are extended, abstractedly from impenetrability and inertia; the object of geometry, therefore, is a notion much more general than that of body, as it comprehends, not only bodies, but all things simply extended, without impenetrability, if any such there be. Hence it follows that all the properties deduced in geometry from the notion of extension must likewise take place in bodies, inasmuch as they are extended . . . There are however philosophers, particularly among our contemporaries, who boldly deny that the properties applicable to extension in general, that is, according as we consider them in geometry, take place in bodies really existing. They allege that geometrical extension is an abstract being, from the properties of which it is impossible to draw any conclusion with respect to real objects; thus, when I have demonstrated that the three angles of a triangle are together equal to two right angles, this is a property belonging only to an abstract triangle, and not at all to one really existing. But these philosophers are not aware of the perplexing consequences which naturally result from the difference which they establish between objects formed by abstraction and real objects, and if it were not permitted to conclude from the first to the last, no conclusion, and no reasoning whatever, could subsist, as we always conclude from general notions to particular." L. Euler, *Letters to a German Princess,* Vol. II, p. 31, Brewster's translation (New York: J. & J. Harper, 1833). From other writings of Euler we may infer that "the object" of geometry is the absolute space required by Newtonian mechanics. Cf. Euler's *Mechanik,* (Greifswold: Koch, 1848–53), Vol. I, p. 9 ff. and Vol. III, p. 3 ff; see also his "Réflexions sur l'espace et le temps," *Histoire de l'Académie des Sciences et Belles Lettres* (Berlin, 1748), p. 328.

How relatively recent are the contemporary views on mathematics may be seen from the fact that a man as learned in the sciences as William Whewell could still maintain as late as 1858, the Kantian doctrine that the axioms of geometry "flow from the Idea of Space," understanding by Space a "form by which knowledge derived from our sensations is moulded." *History of Scientific Ideas* (London, 1847), Vol. I, p. 88 ff. On the other hand, neither in Euclid nor in Archimedes does the equivalent of the term "space" occur. The view that the subject matter of geometry is *space* seems to have been developed under the influence of Platonic metaphysics in the sixteenth century. Cf. E. W. Strong, *Procedures and Metaphysics* (Berkeley: University of California Press, 1936).

2. T. L. Heath, *The Thirteen Books of Euclid's Elements,* Vol. I (Cambridge: Cambridge University Press, 1926), p. 10 ff.
3. For a discussion of some of the problems in perspective relations which faced the Renaissance artists see Erwin Panofsky, "Die Perspective als 'Symbolische Form,' " in *Bibliothek Warburg, Vorträge* (Leipzig: B. G. Teubner, 1924–25), p. 238; also E. Papperitz, "Darstellende Geometrie," *Encyklopädie der mathematischen Wissenschaften,* Bd. III A B 6.
4. Monge begins his *Géométrie Descriptive* with a statement of the aims of the work. The first one stated is to free the French nation from its dependence upon foreign industry by educating the young artisans to handle all sorts of instruments and thus introduce precision into their work. The specific objective which Monge sets for himself is the formulation of a uniform method for representing exactly three-dimensional objects upon a plane, and for deducing from such "exact descriptions of bodies" everything requisite concerning the form and mutual positions of the bodies. "In this way one contributed to the education of the nation by familiarizing young engineers with the application of descriptive arts and by making use of this geometry in order to determine the elements of machines."
5. For detailed accounts see M. Chasles, *Aperçu historique sur l'origine et la développement des méthodes en géométrie* (Paris: Gauthier-Villars et fils, 1889), pp. 189 ff; and G. Loria, "Perspektive und Darstellende Geometrie" in M. Cantor's *Geschichte der Mathematik* (Leipzig: B. G. Teubner, 1898), Vol. IV, pp. 577–637.
6. *Annales de Mathématiques* 16 (1826), p. 209.
7. K. C. G. Von Staudt, *Beiträge zur Geometrie der Lage* (Nürnberg, 1856–1860), pp. 166 ff.
8. F. Enriques, *Vorlesungen über Projektive Geometrie,* 2nd ed. (Leipzig: B. G. Teubner, 1915), p. 340. Chasles, p. 76.
9. H. Hankel, *Die Elemente der Projektivischen Geometrie* (Leipzig: B. G. Teubner,

1875), p. 9. Chasles makes much of the fact that the visual construction of figures was dispensed with, and mentions the tradition that Monge lectured without diagrams, using his hands simply to gesticulate (p. 209).

10. It seems that he was stimulated to address himself to this problem by the fact that Gergonne gave a solution to the famous problem of Appolonius—to draw a circle tangent to three given circles—with the help of algebraic methods. Gergonne had concluded that his success established the superior excellence of the latter over synthetic geometry. Poncelet subsequently gave a geometrical solution.
11. J. V. Poncelet, *Applications d'analyse et de géométrie* (Paris: Mallet-Bachelier, 1864), Vol. II, p. 531. Although this volume was first published in 1862, Poncelet said that he had written it in 1818–19.
12. J. V. Poncelet, *Traité des proprietés projectives de figures* (2nd. ed.) (Paris: Gauthier-Villars, 1865), Vol. I, pp. xi-xii. The first edition of this work appeared in 1822, and the contents of the first volume are unchanged in the second edition.
13. *Applications,* p. 531.
14. That is, the method in which, supposing the problem solved, the sufficient conditions for the solution are derived and located in the data. Cf. J. D. Gergonne's "De l'analise et de la synthèse, dans les sciences mathématiques," *Annales de Mathématique* 7 (1816), p. 348 ff.
15. *Traité,* Vol. I, p. xii.
16. Ibid., p. xiii.
17. Ibid., p. xiii.
18. Carnot had introduced the use of positive and negative signs into geometry to indicate differences in position and direction; see Hankel, p. 13.
19. *Traité,* Vol. I, pp. xii–xiii.
20. Chas. Taylor, *Introduction to the Ancient and Modern Geometry of Conics* (Cambridge, England: G. Bell & Sons, 1881), p. lviii.
21. *Traité,* p. xiv.
22. Thus, when two lines in a plane are said to intersect one another, special cases of intersection are: intersection at right angles (or some other determinate degree) or being parallel to one another. The "general" position of two lines (whether straight or curved) is that of a simple intersection, even if they should happen to be so placed as not to intersect "actually," in contradistinction to the lines being tangent, parallel, or asymptotic to one another—unless, of course, certain other conditions should be specified which the lines must satisfy. *Traité,* p. xv.
23. *Traité,* p. xv.
24. Ibid. Poncelet distinguished three types of correlations between two figures when one of them is obtained from the other by a "general transformation." There is a "direct" correlation between the figures, if they are composed of the same number of parts similarly placed, the sole difference being the absolute size of the corresponding parts. The correlation is "indirect" or "inverse," when the parts of the correlative figures are in a different order, differently placed, though the general relations between the corresponding parts are the same. Finally, the correlation is "ideal" when one or more of the "real" parts of the original figure correspond to imaginary or "ideal" parts in its correlate—that is, when "certain distances and points cease to exist in a geometrical manner." *Applications,* Vol. II, p. 301. For each type of correlation there must be an invariance of certain abstract relations stipulated in the initial conditions for the configuration, so that each figure in the series of figures obtained by "gradual transformations" must be an instance of these relations. That is why the principle of continuity was called by Poncelet "le principe de la con-

tinuité ou permanence des relations mathématiques de la grandeur figurée." Ibid., pp. 319, 337–38.

25. *Applications,* Vol. II, p. 547.
26. Cf. E. Nagel, " 'Impossible Numbers': A Chapter in the History of Modern Logic," in *Columbia Studies in the History of Ideas,* Vol. III. (Reprinted as Chapter VIII in the present volume.) Hankel formulated his principle as follows: "The principle of permanence of formal laws . . . consists in this: if two forms expressed in the general signs of universal arithmetic are equal to one another, then they are to remain equal to one another even when the signs no longer designate simple magnitudes, so that the operations upon them also acquire a different meaning . . . Pure formal mathematics does not consist in a generalization of common arithmetic. It is a completely new science, whose laws are not demonstrated by the latter, but only exemplified, if the formal operations, when applied to actual numbers, yield the same results as the intuitive operations of common arithmetic. In the latter, the definitions of the operations determine their rules of combination, in the former the rules fix the meaning of the operations—or, in other words, they indicate their interpretation and their use." M. Hankel, *Theorie der complexen Zahlensysteme* (Leipzig: Leopold Voss, 1867), p. 11–12.
27. *Applications,* Vol. II, pp. 320 ff., esp. p. 336.
28. Ibid., pp. 327, 330.
29. Ibid., p. 344.
30. Ibid., p. 532.
31. *Annales de Mathématiques* 11 (1820), reprinted in Poncelet's *Applications,* Vol. II, p. 557.
32. Cf. the comments of E. Kötter, "Die Entwicklung der Synthetischen Geometrie," *Jahrbuch der Deutschen Mathematischen-Vereinigung,* Bd. V (1901), p. 122.
33. H. Hankel, *Die Elemente der projectivischen Geometrie* (Leipzig: B. G. Teubner, 1875), p. 9.
34. As Bertrand Russell once remarked, though at a time when still handicapped by the view that pure geometry is the *a priori* science of space: "If the quantities with which we end are capable of spatial interpretation, then, and only then, our result may be regarded as geometrical. To use geometrical language, in any other case, is only a convenient help to the imagination. To speak, for example, of projective properties which refer to the circular points, is a mere *memoria technica* for pure algebraical properties; the circular points are not to be found in space, but only in the auxiliary quantities by which geometrical equations are transformed. That no contradictions arise from the geometrical interpretation of imaginaries, is not wonderful: for they are interpreted solely by the rules of algebra, which we may admit as valid in their application to imaginaries. The perception of space being wholly absent, Algebra rules supreme, and no inconsistency can arise. Whenever, for a moment, we allow our ordinary spatial notions to intrude, the grossest absurdities arise—every one can see that a circle, being a closed curve, cannot go to infinity." *Foundations of Geometry* (Cambridge: Cambridge University Press, 1897), p. 45.
35. Chasles illustrates some of the contemporary insights, confusions, and indecisions concerning the matter at issue. He thought that while Poncelet's Principle "justified" the type of proofs offered by Monge and his school for projective theorems, nevertheless the Principle had not been established rigorously, and could be regarded therefore simply as a good inductive rule. (*Aperçu,* p. 199). Here he was on common ground with Cauchy. But he also thought such proofs could be defended by arguments based on an examination of the general processes of analysis. He distin-

guished between two types of geometrical configurations subject to general conditions: in the first, certain parts and positions of the figure, upon which the construction of the figure does not however depend, are "real" and "palpable"; in the second, these parts no longer occur, so that with respect to the first case they have become imaginary, even though the general conditions for constructing the figure have remained the same. Now the theorems enunciated refer to the "permanent parts" of the figure, parts which are required for the general construction of the figure and which are "real and palpable" in both cases. Hence the theorems are independent of the "contingent parts" of the figure, which may be real or imaginary without altering the construction of the figure. He therefore concluded that if a theorem referring to the "permanent parts" is established for one of the kinds of configurations, "no matter how," it holds equally well for the other kind. (Ibid., pp. 200–1). This was the basis for Chasles' Principle of Continent Relations.

Nevertheless, he was not completely satisfied with the arguments which he offered in support of the Principle, and hoped that some day it would be based on some "metaphysical principle of extension, involving ideas of homogeneity, such as those sometimes employed in the natural sciences." (Ibid., p. 204). On occasions, however, he was aware of the postulational character of his principle and he then wished to reconstruct geometry without having to introduce, in what seemed like an arbitrary manner, just those elements which would make possible the use of general methods of proof. Thus he declared that the principle "gives a satisfactory explanation of the word *imaginary,* employed in pure geometry, where it expresses an *ens rationis* without existence, but which one may regard as having certain properties and which are used as auxiliaries, the reasoning upon them being just as upon a real and palpable object." (Ibid., p. 207).

But he also remarked that "the idea of *imaginaries* would be meaningless if it were not always accompanied by the idea of the real existence of the same object to which it applies . . . We can easily avoid the use of imaginaries in reasoning. It is sufficient to suppose that in addition to the figure for which a certain property is to be proved there exists a second figure, which has the same general character as the first, but which is so constructed that the contingent parts which are imaginary in the original figure are real in the second. . . . We may therefore say that the word "imaginary" is used as a short-hand method of expressing oneself, and that it signifies that reasoning is applied to another general position of the figure in which the parts upon which we reason really exist instead of being imaginary as in the original figure." (Ibid., pp. 368 ff.)

No clear view emerges from Chasles' account of the status of imaginaries. He oscillated in his views between regarding his principle as a method of valid inference, as a device for constructing a formal calculus, and as a suggestion for defining new elements in terms of certain relations between the familiar and traditional elements of the science.

36. J. D. Gergonne, "Essai sur le théorie des définitions," *Annales de. Mathématiques* 9 (1818), p. 6.
37. Gergonne admits there is some truth in Condillac's dictum that science is nothing but a well-constructed language. "It is thus true to say that in perfecting a *science* there is also perfected or created a *language*. Just as a good symbolism in algebra facilitates the work, so a good language, while it does not by itself constitute a science, facilitates its progress." Ibid., p. 11.
38. Ibid., p. 13.
39. Ibid., p. 27.
40. Ibid., p. 23.

41. I.e., what would would today be called vectors.
42. H Grassman, "Die Ausdehnungslehre von 1844," in Grassmann's *Gesammelte mathematische und physikalische Werke* (Leipzig: B. G. Teubner, 1894), Bd. I, Erster Teil, p. 7.
43. Ibid., p. 10.
44. Ibid., p. 22. Italics are not in the text.
45. Ibid., p. 65. The mathematical aspect of logic was worked out by Grassmann and his brother Robert, in the latter's *Formenlehre oder Mathematik* (Stettin: Verlag von R. Grassman, 1872).
46. Ibid., pp. 23–24. Although Grassmann stresses the fact that an intuition of space is required for geometry, he does not maintain that this intuition arises from consiering *objects* in space. The intuition is alleged to be of a unique and fundamental kind: "it is given to us simultaneously with the functioning of our senses with respect to the sensible world."
47. Grassmann gives a very unclear account of the mode of "generation" of continuous and discrete forms. The former is "generated" by a "simple act of creation," the latter by a two-fold act of "postulation and junction." His explanation of the simple act of creation is in effect a psychogenetic but *a priori* derivation of the abstract notion of continuity. However, in either case two fundamental modes of intellectual operation are employed, that of noting two postulated elements as *different* and that of noting them as *equal*. Grassmann argues that by combining each of these operations with the preceding two modes of generation, four genera of forms are obtained: the general doctrine of discrete forms is divided into the two sciences of intensive magnitude (function theory, the differential calculus) and of extensive magnitude. (Ibid., pp. 24–26). This latter discipline contains the foundations of the *Ausdehnungslehre*. On Grassmann's view, therefore, the general theory of forms is competent to develop all branches of pure mathematics, using no other intellectual operations than those which are subsumed under the general or abstract notions of equality, difference, junction, and specification (Sonderung). (Ibid., p. 34).
48. Ibid., p. 35 ff.
49. Ibid., p. 47. It is true that Grassmann sometimes makes concessions to readers who demand an intuitive content for terms which are only implicitly defined. Thus he employs on occasion the *language* of geometry, and talks of points instead of elements, and of translations and rotations instead of operations. Nevertheless, he carefully underscores the point that nowhere does the *Ausdehnungslehre* depend upon the theorems of geometry or upon spatial intuition.
50. Ibid., p. 48.
51. Ibid., p. 297. Grassmann writes as if the *Ausdehnungslehre* contained the most general abstract framework which could be devised for the "concrete" results of geometrical study considered as the science of space. On this point he was mistaken, for the framework which he did constuct would today be regarded as constituting no more than the theory of linear transformations in a multiply extended manifold. Grassmann thus studied simply a special group of transformations—the projective and affine groups, later supplemented by the group of rotations and Euclidean motions. Cf. the remarks of Grassmann's editor, Fr. Engels, Ibid., p. 406.
52. It has therefore close affinities to the contributions of A. F. Moebius, Sir Wm. R. Hamilton, George Peacock, Augustus DeMorgan, and many others. Hankel's *Theorie der complexen Zahlensystems* (Leipzig, 1867) and especially A. N. Whitehead's *Universal Algebra* (Cambridge, England, 1898) contain a systematic presentation of Grassmann's work from this point of view.
53. Grassmann's *Gesammelte mathematische und physikalische Werke,* Bd. III, 2.

Teil, p. 101. Moebius had declined Grassmann's request to write a notice of the *Ausdehnungslehre* because he claimed incompetence in philosophy. Drobisch, to whom Moebius appealed to undertake the task, also refused. Baltzer, subsequently editor of Moebius' collected works, wrote to the latter: "It is impossible for me to enter into his (Grassmann's) ideas—my head becomes dizzy and everything appears sky-blue when I try reading him." Ibid., p. 102. No notice of the *Ausdehnungslehre* appeared in any mathematical periodical except one written by Grassmann himself.

54. K. G. C. von Staudt, *Beiträge zur Geometrie der Lage* (Nürnberg, 1856), pp. iii–iv.
55. What hindered Von Staudt from giving such an interpretation sooner than he did was the difficulty of discovering a suitable interpretation for conjugate imaginary points. Consider an involution of points on a line and let A, A' and B, B' be any two pairs of points in the involution. Von Staudt resolved the difficulty by distinguishing between the involution whose "sense" or direction is specified by ABA' and the involution whose sense is $A'BA$. These two involutions were taken by him to correspond to a pair of conjugate imaginary points.

    The statement that a given imaginary point lies on a given real line is translatable on Von Staudt's proposal into the statement that the involution represented by the given imaginary point lies on the given line. Since "imaginary line" and "imaginary plane" represent respectively a pencil of lines and a pencil of planes in involution, to say that an imaginary point lies on an imaginary plane means: either the (real) carrier of the points in involution (represented by the imaginary point) coincides with the (real) carrier of the pencil of planes in involution (represented by the imaginary plane), or the involution represented by the imaginary point is in perspective relation to the involution represented by the imaginary plane. *Beiträge*, pp. 76 ff. See also J. Lüroth, "Das Imaginäre in der Geometrie." *Mathematische Annalen* 8 (1875), especially pp. 151 ff.
56. Hamilton's method of interpreting the theory of complex numbers as the theory of *pairs* of real numbers, where appropriate definitions of addition, multiplication, equality, etc., have been stipulated for such pairs, is the same in principle as Von Staudt's method of interpreting imaginary elements in geometry. However, the technical knowledge requisite for understanding the latter's work is undoubtedly less common than is the training required for a thorough apprehension of Hamilton's researches on this matter.
57. Chasles, *Aperçu*, p. 224.
58. Kötter, p. 34.
59. This is the well-known theorem that $F + V = E + 2$.
60. Gergonne, "Recherche de quelques-unes de lois générales qui régissent les polyèdres," *Annales de Mathématiques*, Tome VI (1825), p. 157.
61. Ibid., p. 161. Gergonne's scheme of placing dual theorems in parallel columns has been adopted by almost all writers ever since.
62. Gergonne, "Considérations philosophiques sur les éléments de la science de l'étendue," *Annales de Mathématiques*, Tome XVI (1826), p. 210. Gergonne's third major essay on duality appeared in Tome XVII of his *Annales*.
63. Ibid., p. 212.
64. Ibid., p. 230. Italics *not* in text. Desargue's theorem for the plane is demonstrated with the aid of what is essentially Poncelet's principle of continuity.
65. *Annales de Mathématiques*, Tome XV, p. 158. The essentials of this theory for the plane may be stated briefly. Let C be any conic section, for example a circle, and let P be any point in the plane. From P draw the two tangents to C, and let the straight line passing through the points of tangency be *L*. Then *P* is said to be the *pole* and *L*

the corresponding *polar*. In this way there can be made to correspond to each point in the plane one and only one line. Again, suppose *L* is any line in the plane and *a* and *b* any two points on *L*. From *a* draw the two tangents to *C* and join the points of tangency by a straight line, and do the same for *b*; let the point of intersection of the lines so obtained be designated by "*P*." Then P is said to be the *pole* of the line *L*, which in turn is the *polar* of *P*. In this way there is made to correspond to each line in the plane one and only one point. The outcome of this series of constructions is that a one-to-one correspondence has been established between the points and lines of the plane. Consequently, if *A* is any rectilinear configuration in the plane, there will correspond to it another configuration *A'*, such that each point in *A* will correspond to a line in A' (the polar of the point), and each line in *A* will correspond to a point in *A'* (the pole of the line). More generally, if a point describes an arbitrary curve in the plane, its polar with respect to any fixed conic C rolls upon another curve, such that the first curve is the envelope of the polars of the points on the second; these curves are the *reciprocal polars* of one another with respect to *C*. Hence, as Poncelet remarked, "substituting for each point, line, and curve of a given figure the polar, the pole, and the reciprocal polar curve respectively, we obtain a new figure, which is so related to the original figure that the projective relations of the one can be translated immediately into similar relations between corresponding elements of the second curve." J. V. Poncelet, *Traité,* Vol. II, pp. 57 ff.

66. *Bulletin des Sciences Mathématiques, Astronomiques, Physiques et Chimiques,* Tome V (1826), pp. 112–15.
67. Cf. O. Veblem and J. W. Young, *Projective Geometry* (Boston: Ginn & Co., 1910), Vol. I, p. 26.
68. Gergonne, "Recherches sur quelques lois générales régissant les lignes et surfaces algébriques de tous les ordres," *Annales de Mathématiques,* Tome XVII (1827), pp. 216–17.
69. Poncelet, *Traité,* Vol. II, p. 61.
70. The bitter controversy over matters of priority that ensued is of interest only in so far as it contrasts Poncelet's approach to the admitted facts of duality with Gergonne's more general conception of the principle involved. In the heat of the controversy Poncelet did not hesitate to try blackening Gergonne's character in order to establish his own allegation that Gergonne was a plagiarist. Poncelet accused the latter of being a materialist, of disbelieving that God had created the universe, and of maintaining that man had descended from the apes. *Traité,* Vol. II (2nd ed.), p. 394.
71. "Réclamation de M. Poncelet; avec des notes par M. Gergonne," *Annales de Mathématiques,* Tome XVIII, p. 141.
72. *Annales de Mathématiques,* Tome XVII, p. 276.
73. Poncelet, *Traité,* Vol. II (2nd ed.) p. 271, and first published in *Bulletin des Sciences,* Tome IX (1828).
74. Cf. Kötter, pp. 163–67.
75. J. Steiner, *Systematische Entwickelungen der Abhängigkeit geometrische Gestalten* (Berlin, 1832), in his *Gesammelte Werke* (Berlin, 1881), Bd. I, p. 234.
76. Chasles, *Aperçu,* p. 224.
77. The "accidental property" in question is that the point common to the polar planes of those points which are co-planar in the first figure, has this latter plane for its own polar. Ibid., pp. 229–30.
78. Chasles refers to them as the transformations under which anharmonic ratios are invariant. Ibid., p. 255.
79. *Aperçu,* p. 258.

80. *Aperçu,* pp. 408–9.
81. *Aperçu*, p. 290.
82. Moebius, like Plücker, approached the study of geometry with the tools of analysis, but his immediate influence upon geometers was much less than that of Plücker. Like Plücker, Moebius discovered the source of dual theorems to lie in the formal similarities between the analytic expressions which represented the corresponding configurations. The technical innovation which led him to this result was the introduction of "homogeneous coordinates" for points. For example, in plane geometry three points are taken as fixed, and three "weights" are associated with every other point in the plane in such a way that every point will be the "center of mass" with respect to the three fixed points. (A. F. Moebius, *Der barycentrische Calcul* (1827) in *Gesammelte Werke,* Bd. I, p. 6). In consequence, the terms of every equation of a line or curve will be "homogeneous" or of the same degree in the point variables; for example, if z, y, z are the "weights" or coordinates of a point in the plane, the equation of a straight line will take the form: $ax + by + cz = o$; and the equation of a conic will be:

$$ax^2 + bxy + cy^2 + dxz + eyz + fz^2 = o.$$

Moebius shows that any relation between the points of a geometrical figure, when the latter is specified by such homogeneous equations in point-variables, or "weights," also holds for every other figure described by the same system of equations, but referred to a *different* set of fixed points. The figures whose properties are thus represented by the same equation are not simply *similar* to one another, as perhaps one might be led to assume, "but stand to one another in a more general relation"; in fact, they are projectively related to one another, so that Moebius was led on to develop the theory of collineations. (Ibid., p. 9 ff). He thus came to recognize the correspondence between theorems about points and theorems about lines which had been independently discovered by the French mathematicians. (Ibid., p. 371). Moebius observed, however, that these dualities do not depend, as the French school thought, on the theory of poles and polars, but that on the contrary they "must be deduced simply from the theory of straight lines," i.e., from the theory of collineations. (Ibid., p. 373).
83. J. Plücker, *Analytische-geometrische Entwicklungen,* Bd. II (1931), p. 262; *Gesammelte wissenschaftliche Abhandlungen,* Bd. I, pp. 179 ff. and 470. See also Kötter, p. 169.
84. *Anal.-geom. Entw.,* Bd. II, p. v.
85. *Ges. wiss. Abhand.,* Bd. I, p. 149.
86. J. Plücker, *System der Geometrie des Raumes* (1846), p. 322.
87. *Ges. wiss. Abhand.,* Bd. I, p. 173.
88. Ibid., p. 242.
89. *Anal.-geom. Entw.,* Bd. I, p. ix.
90. There is an intimate connection between the work of Grassmann and Plücker. The former developed a part of n-dimensional pure geometry, the latter showed how to regard so-called three-dimensional space, for example, as an n-dimensional manifold with respect to certain elements. Thus, a space whose elements are Plücker's line-complexes turns out to have the same formal properties as Grassmann's five-dimensional spaces. And conversely, each n-dimensional space of Grassmann may be interpreted in terms of the usual three-dimensional space when certain appropriate geometrical configurations are taken as the elements. This is possible simply be-

cause of the identity of the formal patterns in the material which each discusses. Hence, as Felix Klein remarked, "dass man, wenn irgendwelche Veränderliche zu deuten sind, wechselnd bald diese bald jene Anschauungsweise gebrauchen kann, das ist das Wesentliche." F. Klein, *Vorlesungen über höhere Geometrie* (Berlin: Verlag von Julius Springer, 1926), p. 121.

91. M. Pasch, *Vorlesungen über Neuere Geometrie,* Zweite Auflage (Berlin: Verlag von Julius Springer, 1926), p. 3. The first edition of the book appeared in 1882. See also *Mathematik am Ursprung* (Leipzig: Felix Meiner Verlag, 1927), p. 13 ff. Pasch is an "old-fashioned" empiricist because he regarded the truth of his nuclear propositions to be certified by alleged facts of antecedent observation. Hence the empirical truth of the theorems is made to rest on the *certitude* which he ascribed to the nuclear propositions said by Pasch to be "evident." (*Neuere Geom.,* p. 92). On this point he was at one with an empiricist like Mill, who also tried to find the ground of certainty of the axioms in their *source* rather than in the verifiable consequences to which they lead.
92. *Neuere Geom.,* pp. 15–16 and 43.
93. Ibid., p. 41.
94. Ibid., p. 65.
95. Pasch established the general principle of duality in a rigorous way. He first deduced two sets of projective theorems from his nuclear propositions, and found that the statements in one set are the duals of statements in the other. He then proved that all other projective theorems are derivable from one or the other of these two sets of propositions; and that if a theorem can be deduced from the first set another theorem can always be deduced from the second set. It follows that such a pair of theorems will be duals of one another. Ibid., pp. 187–90.
96. Ibid., p. 91.
97. M. Pasch, *Mathematik und Logik* (1924), p. 11. (First published in 1915.) "Der mathematische Beweis hat mit dem, was die Stoffwörter bedueten nichts zu tun; er hält sich in letzter Linie nur an die Fügemittel und stellt so einen reinen *Formalismus* vor." *Mathematik am Ursprung,* p. 138; see also pp. 74 ff.
98. *Mathem. u. Logik,* p. 37.
99. Ibid., p. 42 ff.
100. For an account of Peano's work, see P. E. B. Jourdain, "On mathematical logic and the principles of mathematics," *Quarterly Journal of Pure and Applied Mathematics* 43 (1912), p. 284; also G. Veronese, *Grundzüge der Geometrie* (Leipzig: B. G. Teubner, 1894), p. 683.
101. The work of this school is contained in the early volumes of the *Rivista di Matematica,* and Peano's *Formulaire* (Torino: Fratres Bocca, 1895–1908); Pieri's important monographs appear in the *Atti dell' Accademia di Torino* 30 (1895) and 34 (1898), and also in the *Memoire* of the same Academy, 48 (1898).
102. Veronese, pp. xiii and xxi.
103. Thus, Veronese cites as presuppositions of his system such statements as "I think" and enumerates among the "operations" he requires the process of "abstraction." He also gives a list of some of the logical propositions he assumes, such as "The concept A is the concept A" (p. 1 ff).
104. Ibid., p. xxxii.
105. F. Enriques, "Die Prinzipien der Geometrie," *Enzyklopädie der Mathematischen Wissenschaften,* Bd. III, A B I; *Problems of Science* (LaSalle, Illinois: The Open Court Publishing Co., 1914), Ch. 4; *Historic Development of Logic* (New York:

Henry Holt and Co., 1929), pp. 175–79. Enriques' support on this point is of particular interest, because he rejects Poincaré's alleged "nominalistic" views on the relation of pure geometry to "space" and to physical inquiry.

106. D. Hilbert, *Grundlagen der Geometrie,* 7th ed., (Leipzig und Berlin: B. G. Teubner, 1930), p. 2. The first edition appeared in 1897.

107. Ibid., p. 42.

108. If $x, y, z, w,$ are the projective numerical coordinates of four points on a line, their anharmonic ratio is defined as

$$\frac{(x-z)}{(x-w)} : \frac{(y-z)}{(y-w)}.$$

109. F. Klein, "Vergleichende Betrachtungen über neurere geometrische Forschungen," 1872, in *Gesammelte mathematische Abhandlungen* (Berlin: Verlag von Julius Springer, 1921), Bd. I, p. 463. "Group" is a technical term. A class K of elements $a, b, c, \ldots$ is said to form a group with respect to an "operation" * when it satisfies a number of very specific conditions, among which the following is one of the most characteristic: if a is an element of K and b is also an element, then $a{*}b$ is an element. Thus, the class of cardinal numbers forms a group with respect to the operation of addition, but is not a group with respect to the operation of subtraction. By the "principal group" is understood the largest set of transformations with the group property which is compatible with the invariance of the properties studied in the system under consideration.

110. Ibid., p. 471. For example, suppose that *point-pairs* are taken as the elements of a conic instead of single points, as is usually done. Now the totality of point-pairs on a conic may be made to correspond in a one-to-one manner with the totality of lines in a plane; and it follows by considerations similar to those discussed above that the group of linear transformations which correlate point-pairs of the conic with other such point-pairs, can be made to correspond in a reciprocally unique way with the group of linear transformations in the plane which leave the conic invariant,—that is, which correlate point-pairs of the conic with other such point-pairs. If, finally, we recall the connection between the projective geometry of a line and the projective geometry of a conic, as discussed in the body of the text, we may conclude with Klein that the projective geometry of the line is *structurally identical* with the projective geometry of the plane, when a conic is assumed fixed or invariant under the transformations of the plane.

111. F. Klein, "Ueber die sogenannte Nicht-Euklidische Geometrie," in *Ges. math. Abhand.*, pp. 275 ff., 300 ff.

112. Ibid., p. 253.

113. Ibid., p. 381.

114. H. Helmholtz, "Ueber den Ursprung und die Bedeutung der geometrischen Axione" (1870), in *Vorträge und Reden* (Braunschweig: F. Vieweg & Sohn, 1884), Bd. II.

115. Klein, p. 381.

116. Ibid., p. 386. These remarks were first published in 1898, apropos of the awarding of the Lobatchevsky Prize to the third volume of Lie's *Theorie der Transformationsgruppen* (Leipzig: B. G. Teubner, 1888–93).

117. Ibid., p. 396. Klein also attempted to develop a theory of functions which he thought would do greater justice to the vagueness of our spatial intuitions than does the usual theory. He offered an "approximate representation" for a precisely defined function, which turns out to be not a one-dimensional curve but a two-dimensional

strip (*Funktionsstreifen*). And he maintained that a physicist employing function theory is in fact concerned with just such representations. "The element of spatial intuition . . . is not the single point, but the three-dimensional body. We may conceive the body as diminished in size, but we never obtain in this way the intuition of a single point. Similarly, it is impossible to represent a curve exactly; what we can always see is only a body of which two dimensions fall into the background with respect to the third." "Ueber den allgemeinen Funktionsbegriff und dessen Darstellung durch eine willkürliche Kurve," in *Ges. math. Abhand.*, Bd. II, p. 214. This essay was first published in 1873.

118. F. Klein, *Evanston Colloquium Lectures* (1893), reprinted in *Ges. math. Abhand.*, p. 226. Some further remarks are also of great interest. "The ordinary mathematical treatment of any applied science substitutes exact axioms for the approximate results of experience, and deduces from the axioms rigorous mathematical conclusions. In applying this method, it must not be forgotten that mathematical developments transcending the limits of exactness of the science are of no practical value. . . . When the astronomer says that the periods of two planets must be exactly commensurable to admit the possibility of a collision, this holds only abstractly, for their mathematical centers; and it must be remembered that such things as the period, the mass, etc., of a planet cannot be exactly defined, and are changing all the time. Indeed, we have no way of ascertaining whether two astronomical magnitudes are incommensurable or not; we can only inquire whether their ratio can be expressed approximately by two *small* integers" (pp. 229–30).

119. H. Poincaré, *Foundations of Science* (New York: The Science Press, 1921), p. 65.

120. Ibid., p. 235.

121. Cf. Einstein's remarks in his *Geometrie und Erfahrung* (Berlin: Verlag von Julius Springer, 1921), especially pp. 4–9 in which he acknowledges his indebtedness to Poincaré.

122. Poincaré, pp. 83 ff.

123. Ibid., p. 65. Poincaré believed that Euclidean geometry is and would always remain the most convenient system of measurement for physics. "In astronomy 'straight line' means simply 'path of a ray of light.' If therefore negative parallaxes were found, or if it were demonstrated that all parallaxes are superior to a certain limit, two courses would be open to us; we might either renounce Euclidean geometry, or else modify the laws of optics and suppose that light does not travel rigorously in a straight line. It is needless to add that all the world would regard the latter solution as the most advantageous." (p. 81). In this conjecture Poincaré was a bad prophet and relativity physics has proved him wrong.

11. Some Notes on Determinism

1. Brand Blanchard, "The Case for Determinism," in *Determinism and Freedom*, ed. Sidney Hook (New York: New York University Press, 1958).

2. Max Black, "Making Something Happen," in Sidney Hook.
3. Percy W. Bridgman, "Determinism in Modern Science," in Sidney Hook.
4. Alfred Landé, "The Case for Indeterminism," in Sidney Hook.
5. Paul Edwards, "Hard and Soft Determinism," and John Hospers, "What Means this Freedom?," both in Sidney Hook.

12. Teleology Revisited:

A. GOAL-DIRECTED PROCESSES IN BIOLOGY

1. John Dewey, *Logic* (New York: Henry Holt and Company, 1938), p. 19.
2. John Dewey, *Influence of Darwin on Philosophy and Other Essays* (New York: Henry Holt and Company, 1910), p. 9.
3. Cf. George C. Williams, *Adaptation and Natural Selection* (Princeton, N.J.: Princeton University Press, 1966), pp. 8–9.
4. Andrew Woodfield, *Teleology* (Cambridge, England: Cambridge University Press, 1976), p. 104.
5. Ibid., p. 164.
6. Ibid., p. 193.
7. Ibid., p. 195.
8. Ibid., p. 196.
9. Ernst Mayr, "Teleological and Teleonomic, a New Analysis," in R. S. Cohen and M. W. Wartofsky, eds., *Methodological and Historical Essays in the Natural and Social Sciences,* vol. XIV of *Boston Studies in the Philosophy of Science* (Dordrecht-Holland/Boston: D. Reidel Publishing Company, 1974), p. 98.
10. Ibid., p. 98.
11. Ibid., p. 102.
12. Ibid., p. 99.
13. Ibid., p. 103.
14. E. O. Wilson, *Sociobiology* (Cambridge, Mass.: Harvard University Press, 1975), p. 157.
15. Ernst Mayr, p. 104.
16. G. Sommerhoff, *Analytical Biology* (London: Oxford University Press, 1950).
17. Gerd Sommerhoff, *Logic of the Living Brain* (London: John Wiley and Sons, 1974), Ch. 4
18. Walter B. Canon, *The Wisdom of the Body* (New York: W. W. Norton & Company, 1939), Ch. 4.
19. This example is discussed in G. Sommerhoff, *Analytical Biology,* p. 99. An elementary mathematical analysis of an analogous problem is given in John Cox, *Mechanics* (Cambridge, England: Cambridge University Press, 1923), pp. 210–11.
20. Andrew Woodfield, pp. 67–72.
21. Cf. Francisco Ayala, "Biology as an Autonomous Science," *American Scientist* 56 (1968), p. 214.
22. Robert Cummins, "Functional Analysis," *Journal of Philosophy* 72 (1975).
23. Ibid., p. 748.

B. FUNCTIONAL EXPLANATIONS IN BIOLOGY

1. Walter J. Bock and Gerd von Wahlert, "Adaptation and the Form-Function Complex," *Evolution* 19 (1965), p. 274.
2. Ibid., p. 278.
3. Larry Wright, *Teleological Explanations* (Berkeley: University of California Press, 1976).
4. Ibid., p. 22.
5. Ibid., p. 38, 84.
6. Theodosius Dobzhansky, *Evolution, Genetics and Man* (New York: John Wiley & Sons, 1955), p. 130.
7. Ibid., p. 131.
8. Ibid., p. 132. Cf. also F. J. Ayala, "Teleological Explanations in Evolutionary Biology," *Philosophy of Science* 37 (1970).
9. Ernst Mayr, *Populations, Species, and Evolution* (Cambridge, Mass.: Harvard University Press, 1963), p. 119.
10. I. Kant, *Critique of Judgment*, translated by J. H. Bernard (London: Macmillan and Company, 1931), p. 278.
11. Ibid., p. 295.
12. Ibid., p. 279–81.
13. C. D. Broad, *The Mind and Its Place in Nature* (London: Kegan Paul, Trench, Trubner & Co., 1925), p. 82.
14. C. G. Hempel, *Aspects of Scientific Explanation* (New York: The Free Press, 1965), p. 304.
15. Ibid., p. 306.
16. Ibid., p. 310.
17. Ibid., p. 313.
18. Michael Ruse, *The Philosophy of Biology* (London: Hutchinson University Library, 1973), Ch. 9.
19. Ernest Nagel, *The Structure of Science* (New York: Harcourt, Brace & World, 1961), Ch. 12.
20. Ibid., p. 421.
21. Ibid., p. 408, 422.
22. Vernon Grant, *The Origin of Adaptations* (New York: Columbia University Press, 1963), p. 117.
23. Cf. p. 291 above.

Index